AF767396

Multi-scale and Multi-physics Simulation of Fracture and Fatigue

Multi-scale and Multi-physics Simulation of Fracture and Fatigue presents a comprehensive exploration of theoretical and computational methods used in the analysis of fracture and fatigue in materials, effectively bridging length and time scales. It explores both fundamental concepts of fracture mechanics and cutting-edge multi-scale computational modeling.

By focusing on methods to predict material properties, including fracture toughness, based on nanoscale interactions up to macro-scale failure, the book employs physics-based techniques derived from quantum mechanics, molecular dynamics, and mechanics of materials. These approaches enable readers to uncover structure-property relationships within materials and predict material stiffness properties, aligning with the foundational principles of Integrated Computational Materials Engineering (ICME). Through its detailed discussion of multi-scale and multi-physics techniques essential for fatigue and fracture analysis, the book serves as a thorough guide for analyzing fatigue and fracture in both conventional polymers and innovative materials, including nanostructures.

This book is an invaluable resource for material scientists and mechanical engineers with an interest in multi-scale modeling of fracture and fatigue. Additionally, online figure slides are available to complement the content, providing further support for readers.

Multi-scale and Multi-physics Simulation of Fracture and Fatigue

Samit Roy and Sankha Subhra Aditya

CRC Press
Taylor & Francis Group
Boca Raton London New York

CRC Press is an imprint of the
Taylor & Francis Group, an **informa** business

Designed cover image: Samit Roy and Sankha Subhra Aditya

First edition published 2027
by CRC Press
2385 NW Executive Center Drive, Suite 320, Boca Raton FL 33431

and by CRC Press
4 Park Square, Milton Park, Abingdon, Oxon, OX14 4RN

CRC Press is an imprint of Taylor & Francis Group, LLC

© 2027 Samit Roy and Sankha Subhra Aditya

Reasonable efforts have been made to publish reliable data and information, but the author and publisher cannot assume responsibility for the validity of all materials or the consequences of their use. The authors and publishers have attempted to trace the copyright holders of all material reproduced in this publication and apologize to copyright holders if permission to publish in this form has not been obtained. If any copyright material has not been acknowledged please write and let us know so we may rectify in any future reprint.

Except as permitted under U.S. Copyright Law, no part of this book may be reprinted, reproduced, transmitted, or utilized in any form by any electronic, mechanical, or other means, now known or hereafter invented, including photocopying, microfilming, and recording, or in any information storage or retrieval system, without written permission from the publishers.

For permission to photocopy or use material electronically from this work, access www.copyright.com or contact the Copyright Clearance Center, Inc. (CCC), 222 Rosewood Drive, Danvers, MA 01923, 978-750-8400. For works that are not available on CCC please contact mpkbookspermissions@tandf.co.uk

For Product Safety Concerns and Information please contact our EU representative GPSR@taylorandfrancis.com. Taylor & Francis Verlag GmbH, Kaufingerstraße 24, 80331 München, Germany.

Trademark notice: Product or corporate names may be trademarks or registered trademarks and are used only for identification and explanation without intent to infringe.

ISBN: 978-1-032-57728-9 (hbk)
ISBN: 978-1-032-57730-2 (pbk)
ISBN: 978-1-003-44073-4 (ebk)

DOI: 10.1201/9781003440734

Typeset in Times
by SPi Technologies India Pvt Ltd (Straive)

Contents

About the Authors

Samit Roy is the William D. Jordan Endowed Professor in the Department of Aerospace Engineering and Mechanics at the University of Alabama, USA. He holds a Ph.D. in Engineering Science and Mechanics from Virginia Tech. Dr. Roy's research focuses on multi-scale modeling and life-prediction of fiber-reinforced polymer composites, nanostructured reinforcements, and structural health management, including self-healing smart materials. He has authored over 200 peer-reviewed publications. His distinguished honors include Fellow of ASME, Fellow of the American Society for Composites, ASC Outstanding Researcher Award, and ASC/DesTech Award. He served as Division Chair of ASC's Emerging Composite Technologies Technical Division in 2022.

Sankha Subhra Aditya is currently a postdoctoral researcher at the University of Texas, USA. He completed his Ph.D. in Aerospace Engineering and Mechanics at the University of Alabama under the guidance of Prof. Samit Roy. He received his undergraduate education from the Indian Institute of Technology (IIT) Kharagpur in Aerospace Engineering. His research interests encompass a variety of applications, including polymer composites and electronics packaging, with his contributions focusing on multi-scale modeling of materials and structures, as well as theoretical fracture mechanics.

Preface

What is the exact definition of "fracture" at the nanoscale? This is a non-trivial question that requires serious consideration, not only as a fundamental scientific problem but also because the answer should help provide important insights into the physically correct modeling of fracture processes at the nanoscale that will lead to better design of tougher nanomaterials and metamaterials. Therefore, the aim of this book is to describe theoretical and computational methods specifically developed to study fracture and fatigue in materials, bridging length scales and time scales. The contents range from fundamental concepts in fracture mechanics to the state-of-the-art in multi-scale computational modeling of fatigue and fracture. The overarching objective is to be able to predict fracture toughness in materials at the macro-scale using the atomic interactions based on bond potentials, performed in the true spirit of Integrated Computational Materials Engineering (ICME).

The book is organized in a way that makes it very suitable for use as a textbook for a graduate-level course on Computational Modeling of Materials. The first three chapters are designed to introduce students to the basic concepts of fracture mechanics, fatigue failure, finite element method, and molecular dynamics (MD). The subsequent chapters delve deeper into the realm of computational modeling via simulation and prediction of fracture in crystalline and amorphous solids using powerful techniques such as concurrent coupling, the atomistic J-integral, and crack-tip shielding using nanoparticles.

The book focuses on the description of methods to predict material properties such as fracture toughness based on nanoscale interactions all the way to the macro-scale failure. In recent years, the prediction of stiffness properties such as elastic modulus has been made possible by physics-based, first-principles techniques in quantum mechanics, MD, and self-consistent methods in the mechanics of materials. These techniques are allowing researchers to efficiently describe process-structure-property relationships in materials and predict material stiffness properties in accordance with the basic premise of ICME. However, predicting failure properties, such as fracture toughness of materials, remains somewhat elusive due to the inherently multi-scale and complex nature of the failure process, as discussed in this book. For example, the finite element method, which is frequently used in failure analysis of materials and structures, is incapable of *predicting* fracture toughness, and hence the toughness data for the material must be supplied as input. As mentioned in the previous paragraph, the objective of this book is to be able to overcome this shortcoming and predict the fracture toughness using atomistic simulations.

Consequently, in order to connect atomistic information to macro-scale properties, there is a need for multi-scale and multi-physics methods, which is the primary objective of this book. Based on the authors' extensive experience in this field, this book describes the relevant multi-scale and multi-physics techniques necessary for fatigue and fracture analysis. The book shows how concepts such as the continuum-based J-integral frequently used in fracture mechanics can be adapted to discrete atomistic domain to predict fracture toughness, while including entropic effects that may arise due to thermal motion of atoms at elevated temperatures that are usually unaccounted for at the continuum scale. The physics behind certain fundamental issues that arise related to fracture at the nanoscale due to length-scale effects, such as the critical crack size and the fully developed process zone size, are elucidated.

Last but not least, the book delivers an extensive review of recent efforts to simulate fracture using multi-scale coupling of MD with continuum mechanics. Challenging areas such as the modeling of amorphous materials and finite temperature dynamics are explored and elucidated. The influence of embedded nanoparticles in improving fracture toughness in an amorphous material via crack-tip shielding is explored using atomistic simulations. The feasibility of extracting fatigue crack-growth parameters using MD simulations to predict fatigue life of a material at the macro-scale is also investigated in this book.

In summary, this book serves to leverage the authors' extensive experience in classical continuum fracture mechanics and in atomistic multi-scale modeling to bridge an important gap in our understanding of multi-scale fatigue and fracture, and to develop methodologies for predicting fatigue and fracture in conventional polymers and novel materials, including nano-structured materials and bi-material interfaces. Its novelty lies in clearly delineating the correct rules and procedures for obtaining the fracture toughness in quasi-brittle materials, which is often overlooked by researchers in this field. Comparisons with actual experimental data for model verification, which is an essential step in computational model validation, are presented where such data are available.

Acknowledgments

The authors gratefully acknowledge the assistance of Mr. Prasanna Rajagopal at the University of Alabama for his help with the preparation of the manuscript. The book would not have been possible without funding support from NASA and the Air Force Office of Scientific Research (USA) for various research projects carried out by the authors.

1 Introduction to Fracture Mechanics and Fatigue in Materials at Different Length Scales

1.1 INTRODUCTION: FRACTURE MECHANICS

The study of fracture mechanics can be traced back to the pioneering work of George Irwin in 1948, and he is sometimes referred to as the "Father of Fracture Mechanics." While strength-based failure prediction works well for predicting failure in a structure that does not have a crack or a flaw, the strength-based approach breaks down in the presence of a sharp crack, such as a welding flaw. This is because, according to the mechanics of materials, a stress singularity exists at the tip of a sharp crack, which would then lead to catastrophic failure of the structure when even the smallest load is applied! However, experience tells us that this is not the case in real life, and hence fracture mechanics was developed to study and explain this phenomenon.

Fracture mechanics may be subdivided into two broad categories. The first, linear elastic fracture mechanics (LEFM), applies to brittle materials such as glass, ceramic, most epoxy type polymers, and certain metals (e.g., cast iron, high-carbon steel). The second, elastic plastic fracture mechanics (EPFM), originally explored by Jim Rice in 1968, covers ductile materials such as aluminum, copper, and certain thermoplastic polymers.

The study of fracture mechanics begins with flaw initiation, i.e., the formation of a sharp crack tip in a preexisting flaw in the structure. Under either static loading or repeated cycles of operation (i.e., fatigue loading), the flaw may begin to grow and reach a critical size, a_c. When further load is applied to a structure containing this critical crack length, there is typically a rapid crack growth leading to catastrophic failure due to fracture in a brittle material by the flat dashed line as shown in Figure 1.1. On the other hand, for a "tougher" material, the crack growth may encounter resistance as depicted in Figure 1.1 by the rising fracture resistance curve (or R-curve) vs. crack length behavior in ductile materials (solid curve).

DOI: 10.1201/9781003440734-1

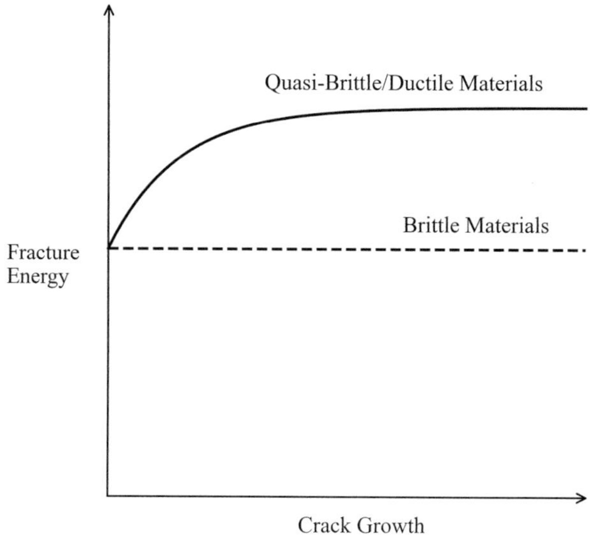

FIGURE 1.1 Rising fracture resistance curve vs. crack length behavior.

1.2 GRIFFITH THEORY OF FRACTURE

Like George Irwin, A.A. Griffith is also considered one of the pioneers of fracture mechanics, especially for his research regarding the study of fracture in brittle glasses [1]. He was investigating the problem that smaller glass objects generally exhibited higher strength than larger glass objects. He postulated that every solid object contains a random distribution of flaws or imperfections due to manufacturing defects, and that failure initiates at the largest of these imperfections. Consequently, larger objects have a greater likelihood of containing larger flaws and will hence fail at a lower applied stress level.

It had been known before Griffith's work that the theoretical fracture strength of glass computed based on atomic bond strength is several orders of magnitude greater than the actual experimentally observed failure strength. Griffith surmised that the large discrepancy was due to the presence of microscopic flaws and defects in the solid that are not present in an ideal solid, and that these cracks could propagate under a load level that is much smaller than the theoretical strength. Griffith adopted an energy balance approach to determine the strength of cracked solids. He postulated that the strain energy released during a crack extension has to be equal to (or greater) than the energy required to create the crack surface. Using this solution, Griffith was able to calculate the total potential energies before and after crack extension. The difference of the potential energies of the two states was set equal to the corresponding gain in surface energy, provided there were no other dissipative mechanisms near the crack tip, such as plasticity.

It follows from the Griffith theory that the fracture strength (the applied stress at the far-field boundaries) of a solid body with a crack is proportional to the square root of the surface energy and is inversely proportional to the square root of the crack size, that is,

$$\sigma_f \propto \sqrt{\frac{\gamma_c E}{a}} \tag{1.1}$$

where, σ_f is the critical applied far-field stress at crack initiation, γ_c is the specific surface energy, a is the half-crack length, and E is Young's modulus. Equation (1.1) clearly explains the reason why the presence of a large-scale flaw would reduce the critical far-field stress, σ_f, well below the theoretical strength in a solid, with larger cracks resulting in lower critical far-field applied stress at fracture. It also demonstrates that the quantity

$$\sigma_f^2 a \propto \gamma_c E = \text{Constant} \tag{1.2}$$

is a material constant, within the framework of LEFM. As mentioned at the beginning of this section, the original work of A.A. Griffith was focused on the fracture of brittle glass. In metals, plastic deformations develop around the crack tip, and the measured fracture strength is much greater than that predicted by the Griffith theory. Orowan and Irwin suggested that the dissipation due to plastic work, γ_p, that occurs in the wake of an elastic-plastic crack should be added to 2γ to define the total work required to create a new crack. For metals, γ_p is significantly larger than the surface energy 2γ, and hence the modified Griffith theory by Orowan and Irwin needs to be used to explain the high fracture strength of metals.

1.3 STRESS INTENSITY FACTOR

To fully understand fracture mechanics, we must first try to understand the behavior of a material in the vicinity of a flaw, which can result in high stress concentration. Figure 1.2 shows a thin plate of width b and length L in the X and Y directions respectively, with an elliptic flaw of length $2a$ and height $2c$ at its center, subjected to a uniaxial far-field stress, σ_0, applied in the Y direction.

However, near the hole, there will be a stress concentration. The highest stress in the Y-direction in this case occurs at point P, as shown in Figure 1.2, and is given by,

$$\sigma_P = \left(1 + \frac{2a}{c}\right)\sigma_0 = K_f \sigma_0 \tag{1.3}$$

In the above Equation (1.1), the stress at point P is an amplification of the far-field stress σ_0 by a factor, $K_f = 1 + \frac{2a}{c}$, which is known as the Stress Concentration Factor (SCF). For an elliptical flaw with an a/c ratio of 10, $K_f = 21$. For a circular hole, $a = c$, resulting in the well-known value of $K_f = 3$. The stress concentration factor slowly reduces to 1 as one moves away from point P along the X-axis, irrespective of the a/c ratio of the flaw.

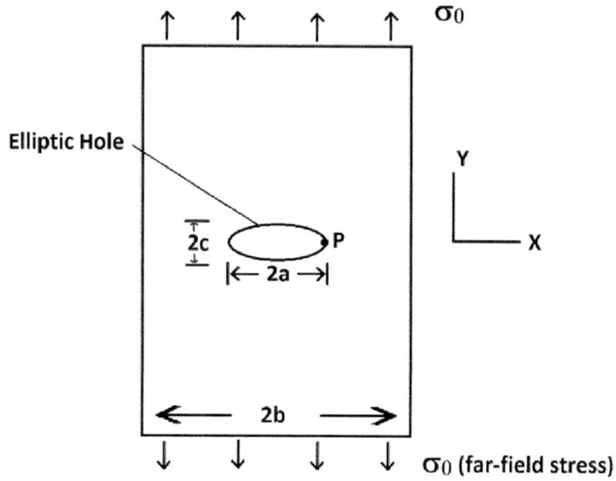

FIGURE 1.2 Plate of width $2b$ with elliptic hole.

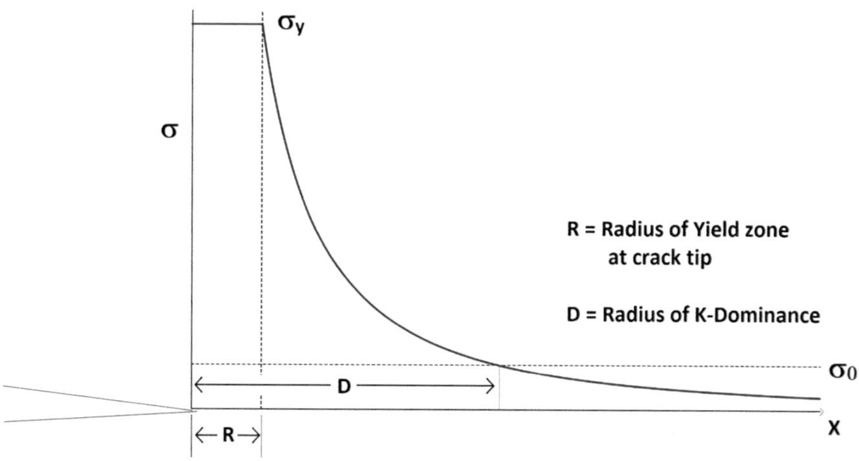

FIGURE 1.3 Stress field near a crack tip showing yield zone and zone of K-dominance.

Now, in Equation (1.1), if we set $c \to 0$, the elliptical flaw becomes a sharp crack. In that case, $K_f \to \infty$. If the stress at point P becomes infinite, then any value of σ_0 should cause the plate to fracture. In reality, however, the stresses at the sharp crack tip do not go to infinity, even for brittle materials. Figure 1.3 depicts the variation of stress as a function of distance from the crack tip. For a material with a yield strength of σ_Y, we find that there is a yield zone of finite radius R near the crack tip where the value of the y-direction stress plateaus at $\sigma_P = \sigma_Y$, even for a so-called "brittle material," leading to what is referred to as small-scale yielding (SSY) near the crack tip. Note that the blue curve in Figure 1.3 intersects the dashed line denoting the far field stress, σ_0, thereby defining the zone of K-dominance, D. For LEFM to be valid for a given material with a flaw, $R/D \ll 1$, as depicted in Figure 1.3, the radius of K-dominance is much larger than the yield radius. For a material with an R/D ratio approaching 1, K-dominance no longer exists, and hence, LEFM is no longer valid.

There are three general modes of loading in fracture mechanics as depicted in Figure 1.4:

1. Mode-I: Opening mode
2. Mode-II: Shearing mode
3. Mode-III: Out of plane shear (tearing) mode

These three modes of loading can occur individually or simultaneously (i.e., mixed-mode loading). For a center-cracked plate subject to Mode-I far-field loading under plane stress conditions, it can be shown that,

$$\sigma_{\text{crack-tip}}(x) = \frac{K_{\text{I}}}{\sqrt{2\pi x}} + \text{Higher order terms} \tag{1.4}$$

where the distance x is measured from the crack tip along the crack plane, that is along $y = 0$. In Equation (1.2), K_{I} is the Mode-I stress intensity factor (SIF), which describes the profile of the stress vs. x curve at the microscale near the crack tip (for $R < x < D$), as depicted in Figure 1.3. For a plate of width $b \to \infty$, we can relate the local SIF, K_{I}, to the far-field applied stress, σ_0, given by $K_{\text{I}} = \sigma_0\sqrt{\pi a}$ for a center-cracked plate of infinite length and width, with a center crack length of $2a$. However, in reality, we need to introduce a geometry correction factor β for a center-cracked plate with finite width b, resulting in the important relation between the local SIF and far-field stress:

$$K_{\text{I}} = \beta\sigma_0\sqrt{\pi a} \tag{1.5}$$

The geometry correction factor β can be readily obtained from any handbook on fracture mechanics. The most important characteristic of the singular term in Equation (1.2) is that the stress field local to the crack tip is identical in form, that is,

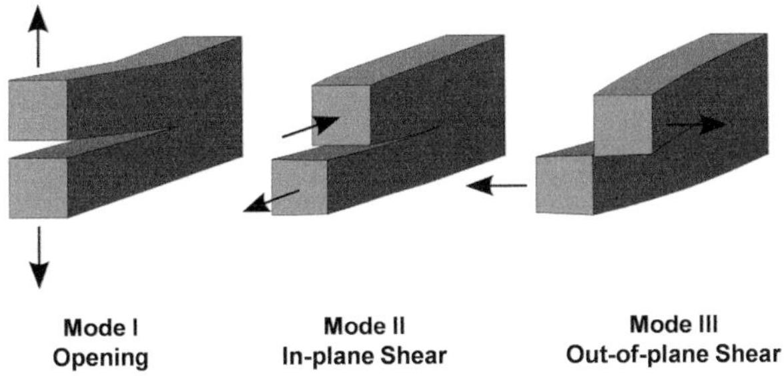

FIGURE 1.4 The three primary modes of fracture.

shows the same $x^{-1/2}$ dependence, for all applied far-field loadings. The far-field loading is only reflected in the local domain by the SIF, K. Therefore, given the crack length and the far-field stress, the local stress field can be fully characterized by the value of K based on Equation (1.3), and, most importantly, a critical value of K can be established as the fracture criterion. Formally, fracture toughness (K_c) is defined as the critical SIF at which the crack will initiate (but not necessarily propagate). For a Mode-I crack to initiate,

$$K_{\mathrm{I}}(\sigma_0, a, \beta) \geq K_{\mathrm{I}c}(T) \tag{1.6}$$

In Equation (1.4), T signifies temperature, implying that fracture toughness is a material property that may be a function of temperature but not of applied loading or crack length. We can invert the relation to find the critical length of the crack at which the crack will initiate for a given applied far-field stress and a given fracture toughness by setting,

$$\beta\sigma_0\sqrt{\pi a_{\mathrm{critical}}} = K_{\mathrm{I}c}$$

giving,

$$a_{\mathrm{critical}} = \frac{K_{\mathrm{I}c}^2}{\beta^2\sigma_0^2\pi} \tag{1.7}$$

Any crack length $a < a_{\mathrm{critical}}$ is then termed a sub-critical crack for the given material and loading condition, and the crack will not initiate under the given far-field load, σ_0. This is an important concept from a structural maintenance point of view, as it allows for the monitoring of cracks in the structure, especially under fatigue loading, until the crack length approaches the critical value, at which point the component must be repaired or replaced.

Similarly, for Mode-II and Mode-III, we have the following relations for critical SIFs:

$$\begin{aligned} K_{\mathrm{II}}(\sigma_0, a, \beta) &\geq K_{\mathrm{II}c}(T) \\ K_{\mathrm{III}}(\sigma_0, a, \beta) &\geq K_{\mathrm{III}c}(T) \end{aligned} \tag{1.8}$$

The units of the SIF are $\mathrm{MPa}\sqrt{\mathrm{m}}$ $\left(\text{or ksi}\sqrt{\mathrm{in}}\right)$. For LEFM to be valid, in all three cases, self-*similar crack growth* is assumed to occur, meaning that there is no sudden change in crack direction or crack bifurcation or crack bridging.

In the presence of combined loading the crack tip experiences what is known as "mixed-mode" fracture. A commonly used mixed-mode failure criterion employed is the power-law criterion [2] given by Equation (1.7)

$$\left(\frac{K_{\mathrm{I}}}{K_{\mathrm{I}c}}\right)^m + \left(\frac{K_{\mathrm{II}}}{K_{\mathrm{II}c}}\right)^n = 1 \tag{1.9}$$

where K_{I} and K_{II} are SIFs in Mode-I and Mode-II respectively, K_{IC} and $K_{\mathrm{II}c}$ are their critical values, and m and n are power-law constants that are material dependent. Another important fracture parameter that relates to the SIF is the strain energy release rate (SERR). For isotropic materials, the Mode-I SERR, G_{I} relates to the K_{I} as follows:

$$G_{\mathrm{I}} = \frac{K_{\mathrm{I}}^2}{E^*} \tag{1.10}$$

where $E^* =$ Young's modulus (E) for plane stress, and $E^* = \frac{E}{1-\nu^2}$, for plane strain, and ν is the Poisson's ratio of the material. The unit for the SERR is energy released per unit area of crack extension (e.g., J/m^2) and it defines the strain energy released per unit area of crack growth. Expressions similar to Equation (1.8) exist for G_{II} and G_{III}, that is, Mode-II and Mode-III SERR, in terms of SIF K_{II} and K_{III} [1], respectively.

1.4 FRACTURE IN ANISOTROPIC MATERIALS

For fracture in anisotropic materials, such as fiber-reinforced composite materials, the definition of SIF in Equation (1.3) remains unchanged. However, the assumption of self-similar crack growth is typically violated due to crack deflection/branching and fiber-bridging of a crack in a composite. In these cases, a model that includes a fracture process zone, such as the cohesive zone model, needs to be employed to incorporate quasi-brittle and mixed-mode fracture behavior near the crack tip. Further, the relation between the Mode-I SERR, G_{I}, and K_{I} given by Equation (1.1) is no longer valid, as the anisotropic compliance of the composite material must replace E^*. The exact expression for the relation between the Mode-I SERR, G_{I}, and K_{I} and for other modes for anisotropic materials, such as composites, can be found in Williams [1].

1.5 LENGTH SCALE EFFECTS IN FRACTURE MECHANICS

Extensive studies have recently been carried out by researchers to investigate the effect of length scale on the mechanics of fracture in solids. Gao et al. [3] used Griffith theory and the Dugdale cohesive fracture model to establish that there exists a critical flaw size below which a material becomes notch-insensitive, and therefore, flaw tolerant. Zhang et al. [4] conducted detailed in-situ micro-tensile experiments as well as molecular dynamics (MD) simulations on a monolayer center-cracked graphene sheet with various crack lengths to assess the effect of nano-size crack length on fracture energy. They concluded from their simulations that LEFM is *valid* for graphene because the parameter $\sigma_c\sqrt{a_0} \approx \sqrt{2\gamma E/\pi}$ remained approximately constant, where σ_c is the critical far-field stress at fracture initiation, a_0 is the semi-crack length, γ is the energy per unit area to form a new surface, and E is the Young's modulus of graphene. However, a closer look at their MD simulation data reveals that this observation is only true for overall semi-crack lengths greater than 8 nm.

A significant departure (≈17%) from the LEFM prediction is observed below this 8 nm critical semi-crack length.

Inspired by these observations, the authors of this book [5, 6] simulated failure in a center-cracked single monocrystalline graphene sheet using MD that further established the existence of a critical semi-crack length of 0.154 nm for a 2-D graphene sheet, below which strength-based rupture dominates. Shimada et al. [7] published an article titled "Breakdown of Continuum Fracture Mechanics at the Nanoscale" wherein they concluded that both the SIF and the SERR fail to describe fracture below a critically confined singular field of 2–3 nm, i.e., breakdown of LEFM within the framework of the continuum theory occurs due to the discreteness of the atomic domain near the nanoscale crack tip. They proposed an energy-based theory, called discrete fracture mechanics (DFM), which is based on discretization of the continuum definition of SERR that explicitly accounts for the discrete nature of atoms in a single crystal silicon, and were able to predict the macro-scale critical SERR accurately as long as the specimen width remained above the 40–60 nm range. However, DFM is not ideal for use in amorphous polymer systems due to the absence of self-similar crack growth caused by lattice bond breakage and difficulty in determining the exact increment in the area of the crack surface due to crack extension in a polymer with entangled molecular chains.

On the other hand, Wilson et al. [8] presented a novel numerical method for the determination of continuum SIFs from atomistic simulations using projection of observed displacements from MD onto the set of continuous displacement fields defined by Williams' asymptotic expansion, with the expansion coefficients determining the SIFs. They applied this method to atomistic simulations of fracture in amorphous SiO_2 with a 3 nm crack using a reactive force field and found excellent agreement with experimental values of fracture toughness in silica glass. Notch insensitivity was not observed, although time-dependence and stochastic nature of fracture toughness were identified.

While the aforementioned brittle-fracture-to-strength-based rupture transition and resulting flaw-size tolerance are derived solely from continuum mechanics considerations (e.g., Griffith theory), there is a nanoscale justification for flaw-size tolerance as well. Non-local interactions can provide a non-singular stress field near the fracture tip, leading to a strength-based failure criterion for isotropic [9] and orthotropic materials [10], as various studies have pointed out. Because MD simulations are intrinsically non-local, limited stress concentration ahead of the crack-tip is likely to result in additional anomalous fracture behavior that deviates from the local continuum mechanics-based Griffith theory. This effect can also be included in the proposed J-integral modeling via MD simulations, in which long-range inter-atomic interactions are inherently non-local, as described in the following section.

1.6 DESCRIPTION OF THE ATOMISTIC *J*-INTEGRAL

Modeling particle interactions at the nanoscale necessitates methods that can account for the nonlinear, complex constitutive behavior at this length scale. MD has been shown to be successful in capturing typical fracture events at the nanoscale [5–7]. MD simulations can be thought of as a bridge between the macroscopic material system's deterministic and local behavior and the non-local statistical nature of atomistic interactions. MD allows us to anticipate the time evolution of a system of interacting particles, as well as estimate the physical parameters of the system. MD (as well as Density Functional Theory (DFT)) is now widely used in analyzing atomistic-scale behavior of materials because it can capture complex particle interactions at the molecular level and help us better understand micro-scale phenomena such as energetic and entropic effects at high temperatures, which are typically not considered in continuum-scale modeling. If a sufficient number of atoms are included in the molecular model and the simulation is carried out for a long enough time span, thermodynamic and mechanical parameters can be determined using statistical mechanics utilizing MD, according to the ergodic principle described in Weiner [11].

There are a wide range of material parameters that characterize the fracture toughness of a material, that is, the capability of a material to resist crack growth. Some of these parameters are only applicable to materials with relatively linear mechanical behavior (e.g. SIF, energy release rate), and others are difficult to measure experimentally (e.g. crack-tip opening displacement). One material parameter that is relatively easy to measure experimentally and to calculate computationally and can be applied to materials both with and without significant non-linear behavior is the J-integral. The conventional J-integral is defined as the amount of energy necessary to extend a crack by an incremental length. It can be determined in experimental specimens using the load-displacement response, and in finite element models using a contour integral surrounding the crack tip. In the case of a linear elastic material, the J-integral is the same as the energy release rate (G) and can be easily correlated to the SIF (K). Because of these conveniences, the J-integral has enjoyed great success as a fracture parameter, especially for materials with nonlinear behavior in the fracture process zone (FPZ), and can also be used at bi-material interfaces. Therefore, inspired by earlier work by Jones et al. [12, 13] on a Lennard-Jones (LJ) crystal, the authors of this book have modified the continuum J-integral to study fracture in amorphous polymeric materials at the nanoscale.

1.6.1 Modifications to Continuum *J*

In conventional macro-scale fracture mechanics, as shown in Equation (1.8), the J-integral is used to quantify the crack driving force available from thermo-mechanical loading in the presence of material nonlinearity. Based on pioneering work by Eshelby [14] and more recent work by Jones et al. [12, 13] for crystalline materials, the total J-integral *vector*, $\boldsymbol{J_T}$, in Cartesian coordinates is given by,

$$\begin{aligned}\boldsymbol{J_T} &= \int_{\partial\Omega} \langle \boldsymbol{S} \rangle \cdot \boldsymbol{N} dA = \int_{\partial\Omega} \left[\langle \Psi \rangle \boldsymbol{I} - \langle \boldsymbol{H}^T \boldsymbol{P} \rangle \right] \cdot \boldsymbol{N} dA \\ &= \int_{\partial\Omega} \left[\langle U \rangle \boldsymbol{I} - \langle \boldsymbol{H}^T \boldsymbol{P} \rangle - \langle T\eta \rangle \boldsymbol{I} \right] \cdot \boldsymbol{N} dA \end{aligned} \tag{1.11}$$

where, $\boldsymbol{\Omega}$ is the total material volume enclosed by the J-integral contour, $\boldsymbol{S}=\Psi\boldsymbol{I}-\boldsymbol{H}^T\boldsymbol{P}$ is the Eshelby stress tensor (second-rank tensor), $\Psi = U - T\eta$ is the Helmholtz free energy density (scalar), U is the internal energy density (scalar), $\boldsymbol{H}$ is the displacement gradient tensor, $\boldsymbol{P}$ is the first Piola-Kirchhoff stress tensor, η is the entropy density (scalar), T is the absolute temperature (scalar), $\boldsymbol{N}$ is the outward unit normal vector to the surface $\partial\boldsymbol{\Omega}$ along which contour the J-integral is being evaluated, $\boldsymbol{I}$ is the identity tensor, and A is the surface area of the contour. The vector form of $\boldsymbol{J_T}$ in Equation (1.9) arises from its fundamental definition as the volume integral of the divergence of Eshelby's Energy-Momentum (i.e., stress) tensor [14], and its in-plane Cartesian components (i.e., X_1 and X_2) are related to the conventional Mode-I, Mode-II, and Mode-III components of J as follows: $J_{1T} = J^{\mathrm{I}} + J^{\mathrm{II}} + J^{\mathrm{III}}$, and $J_{2T} = -2\sqrt{\left(J^{\mathrm{I}}J^{\mathrm{II}}\right)}$. Equation (1.9) reduces to Rice's scalar-valued J-integral when only the first component, J_{1T}, is computed by taking the inner product of $\boldsymbol{J_T}$ with only the X-component of the normal vector $\boldsymbol{N}$ along the J-contour as a special case. The advantage of using Eshelby's more fundamental definition for the J-integral is that it allows us to incorporate entropic effects, because free energy is related to internal energy and entropy ($\Psi = U - T\eta$) from continuum thermodynamics. Note that here bold fonts imply vector or tensor quantities, while normal fonts imply scalar-valued parameters. The $\boldsymbol{J_T}$ vector can be partitioned into *energetic* and *entropic* (vector) components as given by,

$$\boldsymbol{J_T} = \int_{\Gamma} \left[\langle U\rangle \boldsymbol{I} - \langle \boldsymbol{H}^T\boldsymbol{P}\rangle\right]\cdot \boldsymbol{N} dA - \int_{\Gamma} [\langle T\eta\rangle \boldsymbol{I}]\cdot \boldsymbol{N} dA = \boldsymbol{J_U} - \boldsymbol{J_\eta} \tag{1.12}$$

where $\boldsymbol{J_T}$, $\boldsymbol{J_U}$, and $\boldsymbol{J_\eta}$ are the *total, energetic, and entropic* (vector) components of the J-integral, respectively. Here it is assumed that the ensemble average $\langle * \rangle$ is approximated by the time average of the quantity over a sufficiently long period of time, based on the ergodic principle [11]. The critical value of $\boldsymbol{J_T}$ at crack initiation (i.e., $\boldsymbol{J_{TC}}$) is related to the critical fracture energy of the material. Consequently, the J-integral can be used as a suitable metric for estimating the crack driving force (J) as well as the critical fracture energy of the material as the crack begins to initiate (J_c), by using the same MD potential to predict both parameters. This is quite unlike continuum methods where the material fracture toughness (or cohesive law) must be pre-defined as an external input parameter in the model to predict failure.

However, for the conventional macroscale definition of the J-integral to be valid at the nanoscale in terms of the continuum stress and displacement fields and their spatial derivatives, it requires the construction of local continuum fields from discrete atomistic data and using this data in the conventional contour integral expression for J, as given by Equation (1.9). One such methodology was proposed by Hardy [15] that allows for the local averaging necessary to obtain the definition of free energy, deformation gradient, and Piola–Kirchhoff stress as local fields (and divergence of fields) and not just as total system averages. The Hardy approach was successfully implemented by Jones et al. [12, 13] to compute the atomistic J-integral, including energetic and entropic contributions, for a Lennard-Jones crystalline material at elevated temperatures. To incorporate entropic contributions at the nanoscale, they used statistical mechanics to define free energy in terms of partition function (Z) of the configurational energy microstate and then employed the Cauchy–Born Rule and local harmonic approximation to obtain the entropic contribution to J.

A similar approach was applied by the authors [5, 6] for modeling fracture in crystalline 2-D graphene sheets using MD simulation. Details regarding the computation of atomistic J-integral (*energetic and entropic*) based on localization functions using Hardy estimates for various materials are presented in Chapter 4.

1.7 INTRODUCTION TO FATIGUE FAILURE IN SOLIDS

The study of fatigue may be conducted in two different ways. The first is the classical approach and makes use of $S - N$ curves, where S stands for applied maximum stress and N for the number of cycles to failure. The second approach is the fracture mechanics-based approach, which makes use of the eponymous Paris Law [16]. A typical fatigue load cycle is shown in Figure 1.5. The applied load oscillates between σ_{min} and σ_{max}. The fatigue stress ratio, often referred to as the R-ratio, is defined as:

$$R = \frac{\sigma_{max}}{\sigma_{min}} \tag{1.13}$$

For fully reversed fatigue loading, $\sigma_{min} = -\sigma_{max}$, and hence $R = -1$. Typical cycle frequency used in fatigue loading is 5 Hz or less, in order to keep dissipative heat effect on the specimen to a minimum.

1.7.1 Paris Law Formulation

The objective of this section is to familiarize the reader with fatigue testing of cracked specimens to evaluate their fatigue performance. The test specimen with an initial pre-crack is fatigue-loaded with maximum and minimum displacements δ_{max} and δ_{min}, respectively, at a frequency of ~5 Hz. The fatigue crack growth is assumed to obey the Paris Law as given in Equation (1.12).

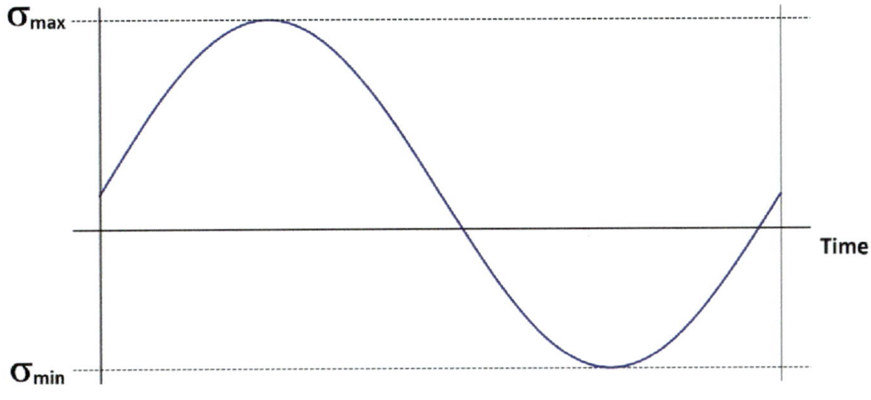

FIGURE 1.5 Typical fatigue load cycle showing a single cycle.

$$da/dN = A\,(\Delta K)^m \tag{1.14}$$

where, a is the fatigue crack length, N is the number of fatigue cycles, ΔK is the applied range of SIF during the fatigue loading for each loading cycle, and A and m are Paris law constants which are material properties. The fatigue crack length a is typically recorded with the help of a digital camera for a particular cycle count, and the corresponding maximum and minimum loads, $P_{\max}$ and $P_{\min}$, are obtained through a load-cell and a data acquisition system, which is subsequently used to calculate ΔK. Accurate measurement of the fatigue crack length, a, is very important and can affect the results considerably. Taking the natural logarithm on both sides of Equation (1.12),

$$\log\{da/dN\} = \log\{A\,(\Delta K)^m\} \tag{1.15}$$

$$\log\{da/dN\} = \log A + m\log\{\Delta K\} \tag{1.16}$$

Once the fatigue testing is completed after running the fatigue tests for about one million cycles (or more if necessary), the fatigue crack growth rate, da/dN, is plotted vs. ΔK, using Equation (1.14) on a log-log scale, and a linear-regression fit is utilized to extract the Paris Law parameters, A and m, given by the intercept and the slope of the curve fit, respectively.

1.7.2 Fatigue Life Prediction in a Solid

Once the Paris Law parameters A and m have been obtained, they can be used to predict fatigue life of a different structure made from the same material. For this purpose, the Paris Law given by Equation (1.12) can be rearranged as follows

$$dN = da/[A\,\{\Delta K(a)\}^m] \tag{1.17}$$

Equation (1.15) can now be integrated to provide the total number of cycles required for the fatigue crack to grow from the initial fatigue delamination length (a_i) to the final fatigue delamination length (a_f) in a stable manner where the initial cycle count is zero, giving:

$$N_f = \int_{a_i}^{a_f} [1/A][1/\{\Delta K(a)\}^m]\,da \tag{1.18}$$

It should be noted that while A and m remain the same for a given material at a given testing temperature, T, the maximum SERR, ΔK can be an entirely different function of crack length, a, in a structure that is different from the original fatigue test specimen.

The basic ideas regarding fatigue analysis presented in this section will be utilized in the simulation of fatigue loading of polymers using MD, as will be discussed in Chapters 3 and 10.

It is important to note that fatigue crack growth presents a number of complications beyond the simple behavior presented by the Paris Law equation. First, the fatigue crack growth must occur in the "stable crack growth" regime for the range of applied ΔK. At one extreme, it is well-established that crack growth will not occur if ΔK is less than some threshold value, ΔK_{th}. On the other hand, a very high ΔK value will lead to very rapid crack growth in the "unstable crack growth regime," and Paris Law is no longer applicable. In addition, there is usually an effect of mean stress in the load cycle.

While the form of Paris Law given in Equation (1.12) is typically used for fatigue in metals, for polymers and polymer composites, a slightly modified form of Paris Law is used,

$$da/dN = A\,(G_{max})^m \tag{1.19}$$

where, G_{max} is the maximum SERR calculated for each loading cycle from the applied maximum load, P_{max}. In general, as the fatigue crack grows with increasing number of cycles, the compliance of test specimen increases, which results in a steady decrease in G_{max}.

1.7.3 A Case Study of Fatigue Behavior in Composite Laminates

For this case study, double cantilever beam (DCB) test specimens were machined from five different unidirectional laminates made of sixteen layers of carbon/epoxy plies, and Mode-I fatigue tests were carried out following ASTM standards in the authors' laboratory. The log (da/dN) vs. log (G_{max}) data thus obtained from the fatigue tests are plotted in Figure 1.6 for DCB specimens from each laminate. The open symbols are the raw test data, whereas the solid lines are the linear regression fit

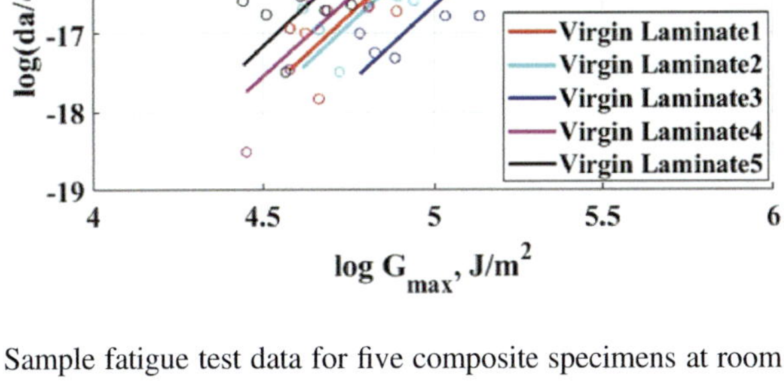

FIGURE 1.6 Sample fatigue test data for five composite specimens at room temperature.

TABLE 1.1
Paris Law Parameters *m* and *A* after Fatigue Testing of Five DCB Specimens

Specimen	m	A
Laminate1	4.06	2.24×10^{-16}
Laminate2	4.06	1.89×10^{-16}
Laminate3	4.01	1.16×10^{-16}
Laminate4	3.95	4.46×10^{-16}
Laminate5	4.1	3.36×10^{-16}
Average	4.04	2.62×10^{-16}
Standard deviation	0.06	1.29×10^{-16}

to the raw data, color coded based on the laminate number. Here, the Paris Law parameters m and A were obtained using a linear fit of the Paris Law as given in Equation 1.17.

It is evident from Figure 1.6 and Table 1.1 that the slope values m for specimens obtained from different laminates are grouped closely together with a low standard deviation, indicating very similar crack growth rates for DCB specimens obtained from each of the five laminates. The values of Paris Law parameters m and A for DCB specimen obtained from laminate1 were 4.06 and 2.24×10^{-16}. Additionally, for virgin DCB specimens obtained from laminate2, laminate3, laminate4, and laminate5, the m values were 4.06, 4.01, 3.95, and 4.1, respectively. The average m value for five specimens was 4.04, with a standard deviation of 0.06. Similarly, the average A value for five specimens was 2.62×10^{-16} with a standard deviation of 1.29×10^{-16}. The differences in the A and m values underscore the variability in composite properties due to slight differences in processing conditions for each laminate, and hence the need to perform a large number of replicate fatigue tests to obtain statistically meaningful results.

1.8 SUMMARY

This chapter provides a brief introduction to the Griffith theory of fracture and the basics of LEFM. The important concept of a critical transition crack length for fracture to rupture transition at the microscale was defined in this chapter. The concepts of SIF and SERR were introduced. The relationship between fracture toughness and critical SERR was described for isotropic and anisotropic materials. The atomistic J-integral and the path independence of J at the nanoscale were also discussed. Last but not least, a detailed discussion of fatigue crack growth in metals and composite materials underscored the need for multiple replicate tests in order to obtain statistically meaningful results in fatigue and fracture. The reader is referred to [2] for a more detailed discussion of advanced topics in linear and nonlinear fracture mechanics.

REFERENCES

1. Williams, J.G., *Fracture mechanics of polymers*. 1984: Ellis Harwood Limited Publisher.
2. Kanninen, M.F. and C.H. Popelar, *Advanced fracture mechanics*. 1985: Oxford University Press.
3. Gao, H. and S. Chen, Flaw tolerance in a thin strip under tension. *Journal of Applied Mechanics*, 2005. **72**(5): p. 732.
4. Zhang, P., Lulu Ma, Feifei Fan, Zhi Zeng, Cheng Peng, Phillip E. Loya, Zheng Liu, Yongji Gong, Jiangnan Zhang, Xingxiang Zhang, Pulickel M. Ajayan, Ting Zhu and Jun Lou, Fracture toughness of graphene. *Nature Communications*, 2014. **5**: p. 3782.
5. Nakarmi, S., T. Sohail and S. Roy, An investigation of flaw-size independence of cohesive traction-separation law at the nanoscale: A molecular dynamics study. *Engineering Fracture Mechanics*, 2019. **215**: p. 235–245.
6. Roy, S. and A. Roy, A computational investigation of length-scale effects in the fracture behavior of a graphene sheet using the atomistic J-integral. *Engineering Fracture Mechanics*, 2019. **207**: p. 165–180.
7. Shimada, T., K. Ouchi, Y. Chihara, and T. Kitamura, Breakdown of continuum fracture mechanics at the nanoscale. *Scientific Reports*, 2015. **5**(1): p. 1–6.
8. Wilson, M.A., S.J. Grutzik, and M. Chandross, Continuum stress intensity factors from atomistic fracture simulations. *Computer Methods in Applied Mechanics and Engineering*, 2019. **354**: p. 732–749.
9. Eringen, A.C. and D.G.B. Edelen, On nonlocal elasticity. *International Journal of Engineering Science*, 1972. **10**(3): p. 233–248.
10. Allegri, G. and F.L. Scarpa, On the asymptotic crack-tip stress fields in nonlocal orthotropic elasticity. *International Journal of Solids and Structures*, 2014. **51**(2): p. 504–515.
11. Weiner, J.H., *Statistical mechanics in elasticity*. 1983: John Wiley & Sons.
12. Jones, R.E. and J.A. Zimmerman, The construction and application of an atomistic J-integral via Hardy estimates of continuum fields. *Journal of the Mechanics and Physics of Solids*, 2010. **58**(9): p. 1318–1337.
13. Jones, R.E., J.A. Zimmerman, J. Oswald, and T. Belytschko, An atomistic J-integral at finite temperature based on Hardy estimates of continuum fields. *Journal of Physics: Condensed Matter*, 2011. **23**: p. 1–12.
14. Eshelby, J.D., The elastic energy-momentum tensor. *Journal of Elasticity*, 1975. **5**: p. 321–335.
15. Hardy, R., Formulas for determining local properties in molecular dynamics simulations: Shock waves. *Journal of Chemical Physics*, 1982. **76**(1): p. 622–628.
16. Paris, P. and F. Erdogan, A critical analysis of crack propagation laws. *Journal of Basic Engineering*, 1963. **85**(4): p. 528–532.

2 An Overview of the Finite Element Method

2.1 INTRODUCTION

The finite element method (FEM) is the mainstay of multi-scale multi-physics simulations in computational mechanics. From the micro scale to the structural scale, FEM provides a versatile framework for modeling mechanical behavior. In its most basic form, FEM assumes a linear elastic material response for linear systems. Outside of the linear regime, a wide variety of geometrically and materially non-linear, inelastic, coupled thermo-elastic, and time-dependent constitutive models have been developed and applied to a range of problems. To model eventual failure, a damage model may be incorporated in FEM, and to model fracture, methods such as the cohesive zone model (CZM), virtual crack closure technique (VCCT), extended finite element method (XFEM), and phase-field models (PFM) are now widely used. However, material parameters that are input to FEM simulations, whether elastic properties such as the Young's modulus or fracture properties such as the critical fracture energy, are acquired either from experimental data or from simulations at lower length scales (e.g., atomistic simulations).

The two primary coupling techniques in multi-scale modeling are the hierarchical and the concurrent simulation schemes. In the hierarchical approach, the physical system is studied in isolation from far-field stimuli and the results are translated to a continuum response using curve fits and/or statistical averaging. One recent example of the hierarchical approach is the Micromechanics Analysis Code with Generalized Method of Cells (MAC/GMC) methodology used by researchers at NASA Glenn Research Center, where the presence of a micromechanical representative volume element (RVE) containing a group of fiber and matrix unit cells as individual constituents in a composite structure is homogenized at the macroscale via its stiffness contribution at individual integration points in an FEM framework, as described in Mital et al. [1]. In this context, Valavala et al. [2] used the energy equivalence of continuum and atomistic models of polymer systems to characterize the nonlinear stress-strain response of polymers (in particular, polycarbonate and polyimide). Buryachenko et al. [3] used the Eshelby and Mori–Tanaka methods to determine the effective properties of nanocomposite materials. Riddick et al. [4] used equivalent modeling of carbon nanotubes (CNTs) in polymers to study the fracture toughness of PMCs with CNTs embedded in them. Awasthi et al. [5] employed molecular dynamics (MD) to determine the force-displacement curves between nano inclusions (CNT) and the polymer system.

In general, it was observed by researchers that the strengthening behavior observed in nanocomposite materials due to nanosized inclusions can only be explained using

DOI: 10.1201/9781003440734-2

theoretical formalisms that appeal to the fundamental physics of interactions at nanometer length scales. True material behavior cannot be accurately simulated through effective or "smeared" continuum analysis since these techniques, by definition, do not explicitly consider the existence of a material interface. Consequently, the "smeared" analyses fail to capture local details, such as favorable surface energy interactions, which are responsible for the stiffness and strength enhancements and failure mechanisms at the nanoscale. It is evident that the stress–strain response of these systems is a function of localized mechanical behavior, and that the strength enhancement is primarily due to attenuation of internal damage resulting from the complex interaction of voids/stress concentrators with the nanosized inclusions in the material. It is for this reason that standard continuum-based multi-scale models, such as the Mori–Tanaka method, can predict linear elastic stiffness accurately, but are unable to predict the material response of nanocomposite systems in the nonlinear regime (damage initiation to failure progression). However, from a design standpoint, it is imperative that we are able to describe the material response from damage initiation to failure progression under all loads and ambient conditions.

Concurrent coupling in multi-scale modeling refers to a method where different models representing different length scales (e.g., atomistic and continuum) are solved simultaneously, with information exchanged between them at each step or iteration. This contrasts with hierarchical coupling, where models with different length scales are solved sequentially one after the other, typically "top down." Advances in computational tools have allowed us to simultaneously model small-scale phenomena using atomistic scale classical techniques such as MD or Monte Carlo (MC) methods, and macroscale responses using continuum level simulations. In MD simulations, the simulation is conducted in a deterministic, time-dependent manner through numerical integration of Newton's equations of motion. In MC, simulations are not time-dependent, but they explore the conformational phase-space of the material system by accepting or rejecting proposed configurations based on a specified algorithm as presented by Wise et al. [6]. However, the challenge in concurrent coupling is to simultaneously process all information available from both length scales and efficiently pass this information continually across time as well as length scales. Concurrent simulations have the unique advantage that they can tie local events at the nanoscale to stimuli from a larger time and length scale during a real-time simulation. One of the earliest forays into concurrent coupling was by Abrahams et al. [7] in their scheme, Multi-scale Atomistic-Continuum Dynamics (MAAD). The scheme involved simultaneous coupling of tight binding (TB), MD, and finite element analysis (FEA). The scheme was successfully applied to study fracture behavior of silicon systems. However, the method had some shortcomings in that it was limited by the time-step size of the tight-binding step (on the order of 10^{-15} seconds) which made the procedure computationally expensive. Ogata et al. [8] used FEM to concurrently couple with density functional theory (DFT) to study the surface oxidation of a silicon system. Some of the more recent efforts in multi-scale concurrent simulations have been the bridging-scale technique, Wagner et al. and Liu et al. [9, 10], and the MD–MPM (Material Point Method) coupling technique proposed by Ma et al. [11]. In Ma et al. [11], the atoms within the handshake region of the MD domain are *directly coupled* pointwise to MPM particles to simulate fracture in a crystalline material. The continuum particles are refined down to atomic sizes, and the boundary material points are assumed to overlap with boundary atoms of the MD zone at all times. The main drawback with this approach is that the thermal vibrations of atoms in the MD domain induce spurious high-frequency vibrations in the continuum domain. Saether et al. [12] proposed an embedded statistical coupling method to overcome the high-frequency vibrations in a concurrent coupling scheme to study grain-boundary decohesion in aluminum. The approach in Saether et al. [12] will be discussed in greater detail in Chapter 5.

In summary, concurrent multi-scale coupling provides a way to blend the efficiency of FEM at the continuum scale with the accuracy of physical models at lower scales. In this chapter, we will explore the numerous ways FEM is used to simulate fracture and fatigue in materials, and how FEM serves as a mathematical framework for multi-scale and multi-physics modeling.

2.2 BASICS OF THE FINITE ELEMENT METHOD (FEM)

2.2.1 Variational Formulation

One of the first steps in developing finite element theory is to begin with the potential energy in a three-dimensional problem of linear elasticity. The total elastic potential energy of the system, $\pi(\boldsymbol{u})$, can be expressed in terms of the stress and strain fields over the volume of the body (Ω), external tractions, and body forces, given by,

$$\pi(\boldsymbol{u}) = \frac{1}{2}\int_{\Omega} \boldsymbol{\epsilon}^{\mathrm{T}} \cdot \boldsymbol{\sigma} d\Omega - \int_{\Omega} \boldsymbol{u} \cdot \boldsymbol{f}_B d\Omega - \int_{\Gamma} \boldsymbol{u} \cdot \boldsymbol{t} d\Gamma \tag{2.1}$$

where "$\cdot$" denotes inner or dot product. In the displacement-based FEM formulation, the independent variable in the above equation is the displacement field vector $\boldsymbol{u}$ from which other quantities, such as stresses and strains, can be derived. At a given loading state on the elastic body, the potential energy must be at a global minimum to achieve stable equilibrium of the structure. Therefore, the first variation of the potential energy, $\delta\pi$, must be zero, as shown below:

$$\delta\pi = \frac{1}{2}\int_{\Omega} \left[\delta\boldsymbol{\epsilon}^{\mathrm{T}} \cdot \boldsymbol{\sigma} + \boldsymbol{\epsilon}^{\mathrm{T}} \cdot \delta\boldsymbol{\sigma}\right] d\Omega - \int_{\Omega} \delta\boldsymbol{u}^{\mathrm{T}} \cdot \boldsymbol{f}_B d\Omega - \int_{\Gamma} \delta\boldsymbol{u}^{\mathrm{T}} \cdot \boldsymbol{t} d\Gamma = 0 \tag{2.2}$$

The above equation can be expressed explicitly in terms of the displacement field using two other relations. Firstly, by assuming a linear elastic form of the material constitutive behavior, the stress tensor can be expressed in terms of the strain tensor at a given point in the body, as shown below,

$$\boldsymbol{\sigma} = \boldsymbol{C}\boldsymbol{\epsilon} \tag{2.3}$$

where $\boldsymbol{C}$ is the elastic stiffness tensor. Next, the strain tensor can be expressed in terms of the displacement field using the strain gradient matrix, as shown below:

$$\epsilon = \begin{Bmatrix} \epsilon_x \\ \epsilon_x \\ \epsilon_x \\ \gamma_{yz} \\ \gamma_{zx} \\ \gamma_{xy} \end{Bmatrix} = \begin{bmatrix} \frac{\partial}{\partial x} & 0 & 0 \\ 0 & \frac{\partial}{\partial y} & 0 \\ 0 & 0 & \frac{\partial}{\partial z} \\ 0 & \frac{\partial}{\partial z} & \frac{\partial}{\partial y} \\ \frac{\partial}{\partial z} & 0 & \frac{\partial}{\partial x} \\ \frac{\partial}{\partial y} & \frac{\partial}{\partial x} & 0 \end{bmatrix} \begin{Bmatrix} u_x \\ u_y \\ u_z \end{Bmatrix} = \boldsymbol{Du} \tag{2.4}$$

By substituting Equations (2.3) and (2.4) in Equation (2.2), the potential energy may now be expressed purely in terms of the displacement fields as follows:

$$\delta\pi = \frac{1}{2}\int_{\Omega}\left[\delta \boldsymbol{u}^{\mathrm{T}}\cdot\left(\boldsymbol{D}^{\mathrm{T}}\boldsymbol{C}\boldsymbol{D}\right)\cdot\boldsymbol{u} + \boldsymbol{u}^{\mathrm{T}}\cdot\left(\boldsymbol{D}^{\mathrm{T}}\boldsymbol{C}\boldsymbol{D}\right)\cdot\delta\boldsymbol{u}\right]d\Omega - \int_{\Omega}\delta\boldsymbol{u}\cdot\boldsymbol{f}_B d\Omega - \int_{\Gamma}\delta\boldsymbol{u}\cdot\boldsymbol{t}d\Gamma = 0 \tag{2.5}$$

Taking advantage of the fact that the constitutive matrix $\boldsymbol{C}$ is symmetric, i.e., $\boldsymbol{C} = \boldsymbol{C}^{\mathrm{T}}$, Equation (2.5) can be simplified as shown below:

$$\delta\pi = \int_{\Omega}\delta\boldsymbol{u}^{\mathrm{T}}\cdot\left[\boldsymbol{D}^{\mathrm{T}}\boldsymbol{C}\boldsymbol{D}\right]\cdot\boldsymbol{u}d\Omega - \int_{\Omega}\delta\boldsymbol{u}^{\mathrm{T}}\cdot\boldsymbol{f}_B d\Omega - \int_{\Gamma}\delta\boldsymbol{u}^{\mathrm{T}}\cdot\boldsymbol{t}d\Gamma = 0 \tag{2.6}$$

At this point, the notion of finite element discretization must be introduced so that the displacement field may be expressed in terms of discrete displacement values at nodal positions within each element. For this purpose, the entire volume of the body (Ω) is subdivided into a number of finite elements. The discretization can best be understood by using the example of a one-dimensional bar under axial loading, as shown in Figure 2.1.

The bar in Figure 2.1 is discretized into three elements of equal length for the purpose of illustration. Each element has a node at each end, and therefore, there

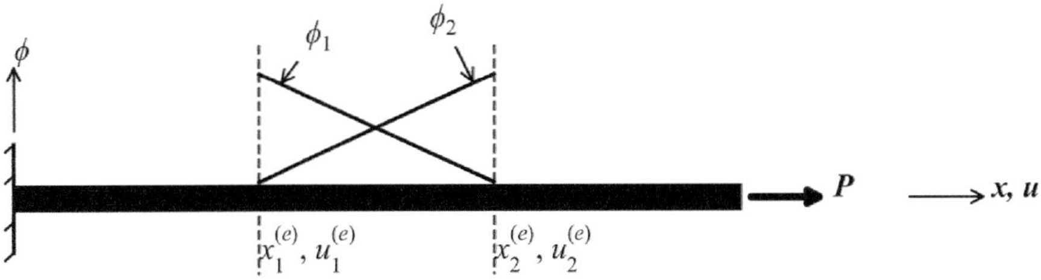

FIGURE 2.1 1-D bar under uniaxial load, discretized into three elements, with linear nodal interpolation functions depicted for an element.

are two nodes per element. The positions of the two nodes of a given element $\left(x_1^{(e)}, x_2^{(e)}\right)$ are shown in Figure 2.1, where the superscript "e" denotes the generic element number, which, in this case, is 2. At each node, we also have the discrete nodal displacements $\left(u_1^{(e)}, u_2^{(e)}\right)$, which are the unknowns. The key to transforming the discrete nodal displacements into continuum displacement fields is through the interpolation function. In Figure 2.1, linear interpolation functions are shown for the two nodes of the given element. For the two-noded elements shown in Figure 2.1, the displacement field interpolated inside a given element can be expressed as shown below:

$$u^{(e)} = \sum_{j=1}^{2} \phi_j u_j^{(e)} \tag{2.7}$$

Interpolation functions have several properties, one of which is that they must be a partition of unity, i.e., $\sum_{i=1}^{2} \phi_i = 1$, in order to avoid spurious stresses occurring due to rigid body motion. Moreover, interpolation functions are usually expressed in terms of local coordinates, as opposed to global coordinates, to facilitate local-to-global element mapping operations. Details regarding these local-to-global coordinate transformations and the Jacobian of the transformation can be found in the book by Reddy [13]. Extending Equation (2.7) to the case of a general element, with any number of nodes and degrees of freedom, the displacement field vector inside a given element number e, $\left(\boldsymbol{u}^{(e)}\right)$ may be expressed as a function of the element nodal displacement vector $\left(\boldsymbol{\Delta}^{(e)}\right)$ and interpolation matrix $(\boldsymbol{\Phi})$ as shown below:

$$\boldsymbol{u}^{(e)} = \boldsymbol{\Phi}\boldsymbol{\Delta}^{(e)} \tag{2.8}$$

Using the above expression for the displacement field within a given element, the strain and stress fields inside the element can be calculated based on Equations (2.3) and (2.4).

Returning to the first variation of the total potential energy (Equation 2.6), the global expression for the entire body can now be replaced with the following equilibrium equation for each finite element:

$$\begin{aligned} \delta\pi^{(e)} = &\int_{\Omega^e} \delta\boldsymbol{\Delta}^{(e)^{\mathrm{T}}} \cdot \left[\boldsymbol{\Phi}^T \boldsymbol{D}^{(e)^{\mathrm{T}}} \boldsymbol{C}\boldsymbol{D}^{(e)}\boldsymbol{\Phi}\right] \cdot \boldsymbol{\Delta}^{(e)} d\Omega^{(e)} - \int_{\Omega^{(e)}} \delta\boldsymbol{\Delta}^{(e)^{\mathrm{T}}} \cdot \left(\boldsymbol{\Phi}^T \boldsymbol{f}_B^{(e)}\right) d\Omega^{(e)} \\ &- \int_{\Gamma^{(e)}} \delta\boldsymbol{\Delta}^{(e)^{\mathrm{T}}} \cdot \left(\boldsymbol{\Phi}^T \boldsymbol{t}^{(e)}\right) d\Gamma^{(e)} = 0 \end{aligned} \tag{2.9}$$

As a first step in simplifying the above equation, the element nodal displacement vectors, which are unknown constants at this point, can be taken outside the integrals, as shown below:

$$\delta\pi^{(e)} = \delta\boldsymbol{\Delta}^{(e)^{\mathrm{T}}} \cdot \left[\int_{\Omega^e} \left[\boldsymbol{\Phi}^{\mathrm{T}} \boldsymbol{D}^{(e)^{\mathrm{T}}} \boldsymbol{C} \boldsymbol{D}^{(e)} \boldsymbol{\Phi} \right] d\Omega^{(e)} \cdot \boldsymbol{\Delta}^{(e)} - \int_{\Omega^{(e)}} \boldsymbol{\Phi}^{\mathrm{T}} \boldsymbol{f}_B d\Omega^{(e)} - \int_{\Gamma^{(e)}} \boldsymbol{\Phi}^{\mathrm{T}} \boldsymbol{t} d\Gamma^{(e)} \right] = 0 \tag{2.10}$$

Now, since $\delta\boldsymbol{\Delta}^{(e)^{\mathrm{T}}}$ is an arbitrary variation in the nodal displacement vector, for $\delta\pi^{(e)}$ to be zero for all values of $\delta\boldsymbol{\Delta}^{(e)^{\mathrm{T}}}$, the expression inside the square brackets must be equal to zero for all values of spatial coordinates within the element e. Therefore, we arrive at the following equality for a given element:

$$\int_{\Omega^e} \left[\boldsymbol{\Phi}^{T} \boldsymbol{D}^{(e)^{\mathrm{T}}} \boldsymbol{C} \boldsymbol{D}^{(e)} \boldsymbol{\Phi} d\Omega^{(e)} \right] \cdot \boldsymbol{\Delta}^{(e)} - \int_{\Omega^{(e)}} \boldsymbol{\Phi}^{\mathrm{T}} \boldsymbol{f}_B d\Omega^{(e)} - \int_{\Gamma^{(e)}} \boldsymbol{\Phi}^{\mathrm{T}} \boldsymbol{t} d\Gamma^{(e)} = 0 \tag{2.11}$$

The first integral on the left-hand side of the above equation is the element stiffness matrix $\left(\boldsymbol{K}^{(e)}\right)$, by way of the strain transformation matrix $\left(\boldsymbol{B}^{(e)} = \boldsymbol{D}^{(e)}\boldsymbol{\Phi}\right)$, and the second and third integrals are the element body-force and surface-force vectors, respectively. The final element-level equation can be written in the compact form:

$$\boldsymbol{K}^{(e)} \cdot \boldsymbol{\Delta}^{(e)} = \boldsymbol{F}^{(e)} \tag{2.12}$$

where,

$$\boldsymbol{K}^{(e)} = \int_{\Omega^e} \boldsymbol{\Phi}^{\mathrm{T}} \boldsymbol{D}^{(e)^{\mathrm{T}}} \boldsymbol{C} \boldsymbol{D}^{(e)} \boldsymbol{\Phi} d\Omega^{(e)} = \int_{\Omega^e} \boldsymbol{B}^{(e)^{\mathrm{T}}} \boldsymbol{C} \boldsymbol{B}^{(e)} d\Omega^{(e)} \tag{2.13}$$

$$\boldsymbol{F}^{(e)} = \int_{\Omega^{(e)}} \boldsymbol{\Phi}^{\mathrm{T}} \boldsymbol{f}_B d\Omega^{(e)} + \int_{\Gamma^{(e)}} \boldsymbol{\Phi}^{\mathrm{T}} \boldsymbol{t} d\Gamma^{(e)} \tag{2.14}$$

Finally, assembly of all the elemental stiffness matrices that constitute the full structure gives us the global stiffness matrix ($\boldsymbol{K}$), and similarly the global force vector ($\boldsymbol{F}$), which, upon applying the suitable boundary conditions can be used to solve for the unknown global displacement vector ($\boldsymbol{\Delta}$). After calculation of the global displacement vector for a given set of loading and boundary conditions, stresses and strains inside each element may be calculated using the strain transformation matrix $\left(\boldsymbol{B}^{(e)}\right)$, given in Equation (2.4), and the constitutive matrix ($\boldsymbol{C}$), given in Equation (2.3), thus completing a static analysis step using FEM.

2.3 ELEMENT TYPES

Lagrange and serendipity elements are two families of elements used in FEM. Serendipity elements do not contain internal nodes and are therefore more computationally efficient compared with corresponding Lagrange elements, without changing the element order. An example of a three-dimensional serendipity element—a 20–noded brick element—is shown in Figure 2.2.

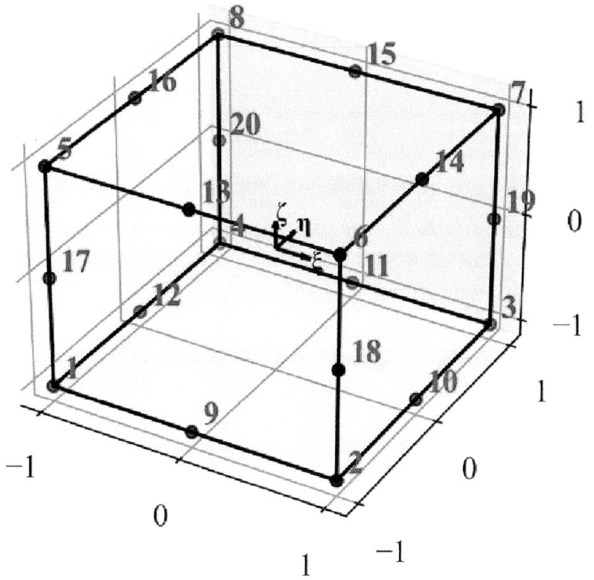

FIGURE 2.2 Schematic of a 20-noded 3D brick serendipity element.

As explained in the previous section, each node of an element has a corresponding interpolation function. Interpolation functions are expressed as functions of the spatial coordinates, with the basic property that they must be equal to one at the corresponding node and zero at all other nodes. Moreover, the sum of all the interpolation points at any point in space must be equal to one, i.e., partition of unity, in order to prevent spurious stresses occurring due to rigid body motion:

$$\sum_{i=1}^{N_{\text{nodes}}} \phi_i = 1 \tag{2.15}$$

It is convenient to use transformed spatial coordinates ranging from −1 to 1, which allows for generalization regardless of element geometry. For example, nodes 1–8 in Figure 2.2 have interpolation functions of the following form:

$$\phi_i(\xi, \eta, \zeta) = \frac{1}{8}(1 + \xi_i \xi)(1 + \eta_i \eta)(1 + \zeta_i \zeta)(\xi_i \xi + \eta_i \eta + \zeta_i \zeta - 2) \tag{2.16}$$

where ξ, η, ζ are the element local coordinates and ξ_i, η_i, ζ_i are the coordinates of any specific node, as depicted in Figure 2.2. The transformation between local and global coordinates takes place using a Jacobian matrix, which must be calculated for each element in the simulation based on its specific geometry.

2.4 SOLUTIONS TO NONLINEAR PROBLEMS

Nonlinearity in FEM problems can arise from two primary sources: geometric nonlinearity, for example, large deformations and moving contacts, and material nonlinearity, for example, nonlinear elastic or elastoplastic constitutive behavior. The general form of nonlinear equations in FEM is as follows:

$$[K(u)]\{u\} = \{F(u)\} \tag{2.17}$$

where K is the global stiffness matrix of the FEM assembly, u is the nodal displacement vector, and F is the global force vector. As can be seen in Equation (2.17), both the stiffness matrix and force vector can be functions of the displacement vector, thereby making the problem fully nonlinear. One approach to the solution of Equation (2.17) is by direct iterations, i.e.,

$$\{u\}_{i+1} = [K(u_i)]^{-1}\{F(u_i)\} \tag{2.18}$$

where u_i and u_{i+1} are the ith and $(i+1)^{\text{th}}$ iterations of the displacement vector, based on an initial guess u_0 for the first iteration. The primary disadvantage of the direct iteration method is that convergence to the correct solution is not guaranteed, and in fact, divergence may occur in some cases. Newton's method is a more reliable approach, in which we first express Equation (2.17) in a residual form, $R(u)$, implying that the Equation (2.19) is not exactly equal to zero for the ith iteration:

$$\{R(u)\} = [K(u)]\{u\} - \{F(u)\} \tag{2.19}$$

By expanding the residual $R(u)$ in the form of a Taylor series about the displacement u_i, gives,

$$\{R(u_{i+1})\} = \{R(u_i)\} + \left[\frac{\partial\{R(u)\}}{\partial\{u\}}\right]_{u=u_i} \left(\{u\}_{i+1} - \{u\}_i\right) + \text{H.O.T.} \tag{2.20}$$

By ignoring the higher-order terms (H.O.T) and forcing the eventual residual to go to zero, i.e., $\{R(u_{i+1})\} = 0$, we get:

$$\{R(u_i)\} + \left[\frac{\partial\{R(u)\}}{\partial\{u\}}\right]_{u=u_i} \left(\{u\}_{i+1} - \{u\}_i\right) = 0 \tag{2.21}$$

Equation 2.21 can be inverted to get an updatable iterative form,

$$\{u\}_{i+1} = \{u\}_i - \left[\frac{\partial\{R(u)\}}{\partial\{u\}}\right]_{u=u_i}^{-1} \{R(u_i)\} \tag{2.22}$$

The derivative of the residual gives

$$\left[\frac{\partial\{R(u)\}}{\partial\{u\}}\right] = [K(u)] + \frac{\partial}{\partial\{u\}}([K(u)])\{u\} - \frac{\partial\{F(u)\}}{\partial\{u\}} \tag{2.23}$$

Using the above form of the residual in Equation (2.23) is known as the exact Newton method. In order to preserve solver symmetry, the last two terms in Equation (2.23) are usually ignored, resulting in,

$$\left[\frac{\partial\{R(u)\}}{\partial\{u\}}\right] \approx [K(u)] \tag{2.24}$$

Equation (2.22) then becomes:

$$\{u\}_{i+1} = \{u\}_i - [K(u)]^{-1}_{u=u_i}\{R(u_i)\} \tag{2.25}$$

Equation 2.25 can be iteratively solved until a given solution convergence criterion such as the L2 norm (Euclidean norm) given by $\sqrt{\frac{\Delta u_i^{\mathrm{T}} \Delta u_i}{u_i^{\mathrm{T}} u_i}}$ decreases to a value below a specified error tolerance. The advantage of Newton's method is that, in general, it provides rapid and stable convergence, as opposed to the direct iteration method, although the convergence depends on the initial guess being within the radius of convergence. A modified Newton method is also useful, wherein the stiffness matrix K is only calculated at the first iteration and not reevaluated at every iteration, which enhances computational efficiency for large systems of equations, although it may slow down the rate of convergence.

2.5 FEM MODELING OF ARBITRARY POINT FORCE INSIDE ELEMENT

In some situations, especially related to concurrent coupling where coupling of the centroid of a finite element to the center of mass of a group of atoms is performed using a harmonic spring, the effect of a point force acting *inside* an element may need to be considered. For example, Figure 2.3 shows a 20-noded three-dimensional brick element with a point force vector $\{F_p\}$ acting at an arbitrary point $P(x_p, y_p, z_p)$ inside the element, due to the force exerted by the MD region via a harmonic spring, as will be discussed in detail in Chapter 5.

Based on the expression for the effective body-force vector for a three-dimensional brick element,

$$\{F\} = \iiint [\Phi]^{\mathrm{T}}\{f_B\}dxdydz \tag{2.26}$$

For a point force acting at point P inside the element, the body force vector may be expressed as a Dirac delta function acting at location $P(x_p, y_p, z_p)$:

$$\{f_B\} = \{F_p\}\cdot\Delta(\{x\} - \{x_p\}) \tag{2.27}$$

where $\Delta(\{x\} - \{x_p\})$ is the Dirac delta function acting at point P. Substituting Equation (2.27) in Equation (2.26), gives,

$$\{F\} = \iiint [\Phi]^{\mathrm{T}}\{F_p\}\cdot\Delta(\{x\} - \{x_p\})dxdydz \tag{2.28}$$

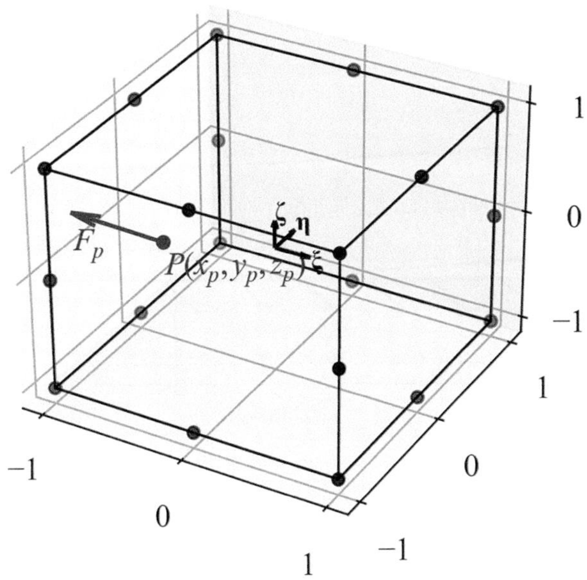

FIGURE 2.3 Schematic of point force acting at a point P inside a 20-noded brick element.

The above equation simplifies, based on the property of the Dirac delta function, to

$$\{F\} = \left[\Phi\left(x_p, y_p, z_p\right)\right]^{\mathrm{T}} \{F_p\} \tag{2.29}$$

In Equation (2.29) the effect of the internal point force at an arbitrary location P inside the element is thus distributed to each of the nodes of the element via the interpolation matrix Φ evaluated at the coordinates of the point force in a weighted manner.

2.6 SIMULATION OF FAILURE AND FRACTURE USING FEM

2.6.1 Cohesive Zone Models

The CZM is now extensively used to model debonding at material interfaces. At the structural scale, CZM is used to model crack growth in adhesive joints, and delamination in layered composite structures. A CZM-based schematic of crack growth in an adhesively bonded joint is depicted in Figure 2.4, showing (a) the cracked, (b) damaged (i.e., process zone), and (c) undamaged (elastic) regions. At the microscale, CZM has been used to model intergranular fracture in polycrystalline materials as discussed in Needleman [14, 15]. Another example of the application of CZM at the microscale is in fiber–matrix debonding in polymer composites to simulate experiments such as the microdroplet test discussed in Tvergaard et al. [16]. The fracture process zone (FPZ) approach of Needleman [14, 15], and Tvergaard et al. [16, 17], involves attributing a traction-separation law to the cohesive interface and, because it

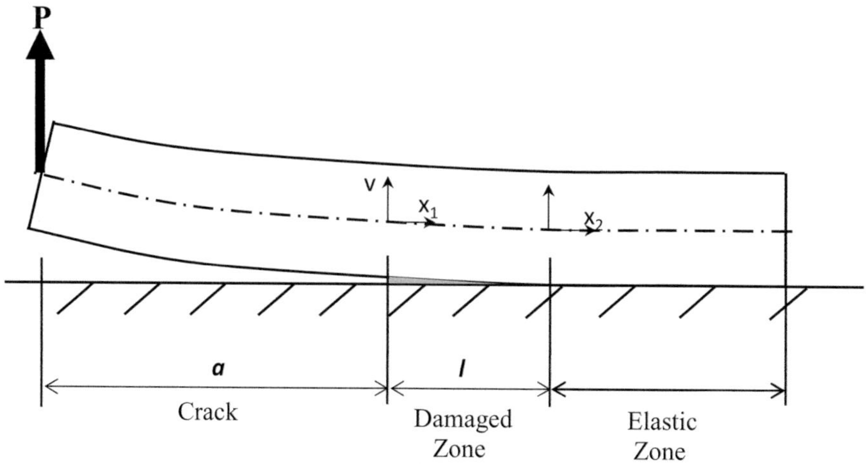

FIGURE 2.4 Schematic of debonding of a bonded joint showing cracked, damaged, and elastic zones.

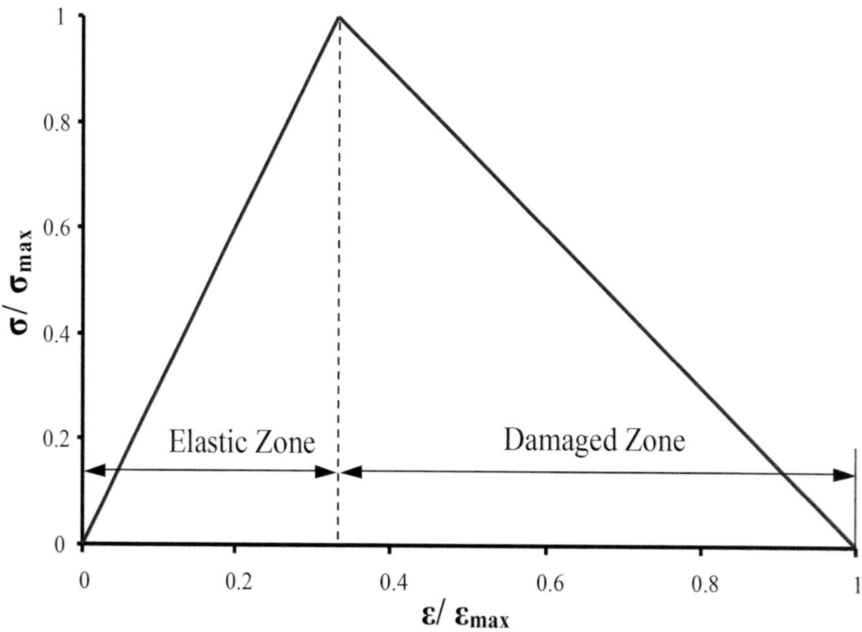

FIGURE 2.5 Triangular traction-separation law (TSL) used in CZM.

allows crack growth to occur, the associated inelastic dissipation from loading and unloading of points that are passed by the crack front is rigorously accounted for. As a result, the selected traction–separation law determines the work of separation (or adhesive fracture energy), which is the work required to create a unit area of a fully developed crack, as mentioned in De Borst [18]. Figure 2.5 shows a typical triangular traction–separation law utilized in most FEM-based CZM analyses, although the

x-axis in Figure 2.5 quantifies normalized strain rather than separation at the crack tip. The CZM is therefore a good tool for investigation of local fracture processes in composite delamination, where inelastic dissipation effects are usually present in the FPZ, particularly when the fracture process is deemed quasi-brittle.

Fatigue crack growth is traditionally characterized via linear elastic fracture mechanics concepts, where crack growth rates are correlated with the change in energy release rate or the maximum value of the energy release rate in a cycle. This approach has worked well for metals and polymers alike, especially in dry, room-temperature environments, where conditions are still generally linearly elastic. Correlation between crack growth rates and elastic fracture parameters does become suspect in polymers near their glass transition and when saturated by a solvent. Cohesive zone modeling offers a solution to this difficulty, in the sense that if the near-tip damage can be accounted for in the traction–separation law of the interphase via homogenization then the global nonlinear inelastic behavior of the material can be coupled into any FEM-based analysis directly as discussed in Knauss et al. [19].

2.6.2 VCCT for Modeling Fracture

The VCCT, first introduced by Rybicki and Kanninen [20], is a widely used LEFM-based approach for computing strain-energy-release rates directly from finite-element nodal forces and displacements near a crack tip. The key idea is that the work required to close a crack extension equals the energy needed to create that extension; thus, mode-partitioned energy-release rates $G_{\mathrm{I}}, G_{\mathrm{II}},$ and G_{III} can be obtained from a single finite element (FE) solution without remeshing. Subsequent developments established robust node- and element-based implementations, clarified mesh-size requirements, and provided mixed-mode separation formulas suitable for shell, plate, and 3-D solid elements. VCCT has seen extensive use for delamination growth in composite laminates (Mode-I, II, and mixed-mode), adhesive joint fracture, and crack propagation in bonded/built-up structures, and it is implemented (natively or via user subroutines) in most commercial FE codes. Comprehensive reviews by Krueger [21, 22] summarize the method's history, best practices (e.g., consistent crack-front forces, appropriate crack-extension length, and mesh bias at the crack tip), accuracy limits (small-scale yielding, coarse meshes, contact at crack faces), and comparisons to alternatives such as CZM and the J-integral methods. However, unlike CZM, VCCT does not explicitly account for the "damaged zone" (aka FPZ) as depicted in Figure 2.4.

2.6.3 XFEM

Another important method for modeling fracture and fatigue is the XFEM. As described in the seminal article by Zi and Belytschko [23], XFEM introduces enrichment functions to complement standard FEM displacement fields so that discontinuities, which are essential to crack initiation and propagation problems, can be handled. XFEM's main advantage is that, unlike CZM and VCCT, the crack path does not have to be known *a priori* in XFEM. However, the computational cost of using XFEM can escalate in the presence of multiple cracks propagating within the analysis domain, and/or crack bifurcation.

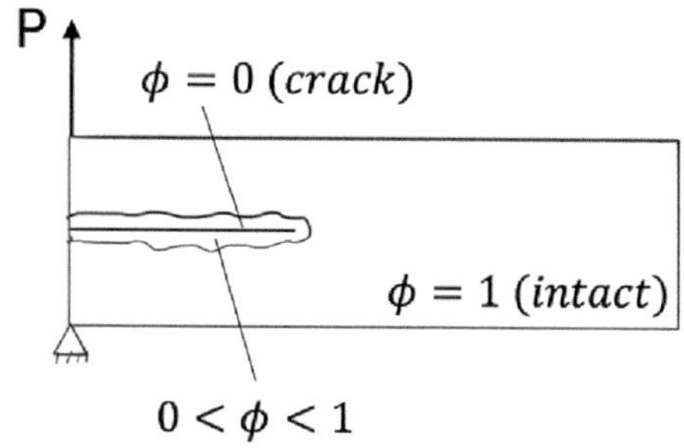

FIGURE 2.6 Phase-field modeling of a cracked beam specimen.

2.6.4 Phase-Field Models (PFMs)

PFM starts from the variational formulation of Griffith's theory [24]. This variational approach treats fracture not as a geometric tracking problem, but as a coupled field evolution problem, where a scalar auxiliary variable (the phase field) smoothly transitions between undamaged and damaged states, as depicted in Figure 2.6 for a cracked beam under an applied load *P*.

The mathematical basis of the PFM is summarized in the following paragraphs. The total potential energy functional can be expressed as,

$$\Psi(u, \Gamma) = \int_{\Omega \backslash \Gamma} \psi_0(\epsilon(u)) d\Omega + \int_{\Gamma} G_c d\Gamma - W_{ext} \tag{2.30}$$

where u is the displacement field, ψ_0 is the strain energy density of the undamaged material, G_c is the critical strain energy release rate (material toughness), and W_{ext} is the work done by external forces.

However, the direct minimization of Ψ is difficult as the area of the crack surface is initially unknown as a function of damage evolution.

However, Equation (2.1) has the form of a Mumford-Shah [25] type functional which is typically found in the field of image processing. Ambrosio and Tortorelli [26] showed that minimization of this type of functional can be approximated by replacing the second surface integral by a volume integral over the domain as:

$$\int_{\Gamma} G_c d\Gamma = \int_{\Omega} G_c \left(\frac{(1-\phi)^2}{4l} + l(\nabla\phi)^2 \right) d\Omega \tag{2.31}$$

here $\phi \in [0, 1]$ is the phase-field (damage) parameter where $\phi = 1$ refers to the undamaged and $\phi = 0$ refers to the fully damaged material, and l is the regularization length scale parameter. The total regularized energy functional becomes,

$$\Psi_l(u, \Gamma) = \int_{\Omega} [g(\phi)\psi^+(\epsilon) + \psi^-(\epsilon)] d\Omega + \int_{\Omega} G_c \left(\frac{(1-\phi)^2}{4l} + l(\nabla\phi)^2 \right) d\Omega \tag{2.32}$$

Bourdin, Francfort, and Marigo adapted this approach for fracture mechanics [27, 28]. The function $g(\phi)$ is the degradation function, which couples the mechanical field to the phase field. A common choice is the quadratic degradation function (e.g., $g(\phi) = \phi^2$). The strain energy is decomposed into tensile (ψ^+) and compressive (ψ^-) modes to address the asymmetry in tensile and compressive response because a crack will not grow in compression. Therefore, the degradation function is only applied to the tensile part. There are two common decomposition methods in the literature: volumetric-deviatoric split [29] and spectral decomposition [30].

Hence, the governing equation for the phase-field approach becomes,

$$\frac{2lg'(\phi)\psi^+}{G_c} + \phi - 4l^2\nabla^2\phi = 1 \tag{2.33}$$

in conjunction with the stress equilibrium equation for a solid body,

$$\nabla \cdot \sigma + b = 0 \tag{2.34}$$

Here stress is defined as the degraded stress,

$$\sigma = g(\phi)\frac{\partial\psi^+(\epsilon)}{\partial\epsilon_{ij}} + \frac{\partial\psi^-(\epsilon)}{\partial\epsilon_{ij}} \tag{2.35}$$

The benefit of this phase-field approach to fracture is that it reduces the complex fracture problem into solving two coupled partial differential equations. The main attributes of this approach are its simplicity and generalizability. Comparisons with other traditional computational fracture modeling approaches are listed below:

1. **CZM:** Perhaps the most severe limitation of the intrinsic CZM is that the crack path must be known *a priori*. Cohesive elements must be inserted beforehand along element boundaries where the crack is expected to grow. If the physical crack path deviates from these pre-defined boundaries (e.g., crack migration in composites), CZM cannot capture it correctly. PFM, on the other hand, allows the crack to propagate in any direction based on the energy gradient, making it superior for "blind" fracture predictions. Additionally, CZM explicitly models the FPZ via the softening curve of the traction separation law. PFM regularizes this zone via the length scale "l". There are studies showing that PFM can be interpreted as a "regularized CZM." By selecting specific degradation functions $g(\phi)$, one can recover specific traction-separation behaviors (e.g., linear softening). This implies that PFM can be tuned to match the material behavior captured by CZM [31–33]
2. **XFEM:** XFEM is very effective for a single, distinct crack. However, it struggles significantly with complex topologies. Modeling crack branching, coalescence, or multiple interacting cracks requires complex algorithms to update the enrichment functions and tracking "level-sets." This often leads to convergence issues in commercial implementations. PFM handles topology changes naturally with no special algorithmic treatment. However, XFEM

allows for a relatively coarse mesh because the singularity is captured mathematically by the shape functions. PFM requires a mesh size h that is a fraction of the regularization length (typically $h < l/2$) to resolve the crack profile. In simple single-crack scenarios, XFEM is significantly faster and requires far fewer degrees of freedom (DOFs). However, as damage complexity increases, PFM becomes more viable despite the cost due to its inherent robustness [34]

PFM has been widely applied to brittle and quasi-brittle materials, such as ceramics, glass, rocks, and concrete, where it accurately captures complex crack patterns under monotonic and dynamic loading.

In ductile materials, the model has been extended by coupling with plasticity, strain hardening, and damage evolution laws, enabling simulation of void nucleation, growth, and coalescence leading to fracture. These formulations have proven effective in predicting failure under large deformations, high triaxiality, and mixed-mode loading [31].

2.7 SUMMARY

The objective of this chapter is to provide the reader with an overview of the finite element method (FEM) and techniques that are relevant to the overarching goal of multiscale failure modeling. Essential concepts that underpin FEM models are first reviewed including the variational formulation, serendipity elements, and nonlinear solution methods. Next, advanced concepts related to FEM based modeling of fracture are introduced including the cohesive zone model (CZM), virtual crack closure technique (VCCT), and extended finite element method (XFEM), and the phase-field model (PFM).

REFERENCES

1. Mital, S.K., S.M. Arnold, P.L.N. Murthy, BA Bednarcyk, E. Pineda, Modeling of time-dependent strength degradation of SiC/SiC ceramic matrix composites via MAC/GMC computer code. NASA/TM-2018-219939. 2018.
2. Valavala, P.K., T.C. Clancy, G.M. Odegard, and T.S. Gates, Nonlinear multiscale modeling of polymer materials. *International Journal of Solids and Structures*, 2007. **44**: p. 1161–1179.
3. Buryachenko, V.A., A. Roy, K. Lafdi, K.L. Anderson and S. Chellapilla, Multi-scale mechanics of nanocomposites including interface: Experimental and numerical investigation, *Composites Science and Technology*, 2005. **65**: p. 2435–2465.
4. Riddick, J.C., S.J.V. Frankland, and T.S. Gates, Multiscale analysis of delamination of carbon fiber-epoxy laminates with carbon nanotubes. in *AIAA Paper, Newport, RI*, May 2006, AIAA-2006-1676. 2006.
5. Awasthi, A.P., D.C. Lagoudas, and D.C. Hammerand, Modeling of interface behavior in carbon nanotube composites. in *AIAA Paper, Newport, RI*, May 2006, AIAA-2006-1677. 2006.
6. Wise, K.E., State of the art assessment of simulation in advanced materials development. NASA-TM-2008-215118. 2008.
7. Abraham, F.F., J.Q. Broughton, N. Bernstein, and E. Kaxiras, Concurrent coupling of length scales: methodology and application. *Physical Review B*, 1999. **60**(4): p. 2391–2403.

8. Ogata, S., E. Lidorikis, F. Shimojo, A. Nakano, P. Vashishta and R.K. Kalia, Hybrid finite-element/molecular-dynamics/electronic-density-functional approach to materials simulations on parallel computers. *Computer Physics Communications*, 2001. **138**: p. 143–154.
9. Wagner, G.J. and W.K. Liu, Coupling of atomistic and continuum simulations using a bridging scale decomposition. *Journal of Computational Physics*, 2003. **190**: p. 249–274.
10. Liu, W.K., E.G. Karpov, and H.S. Park, *Nano mechanics and materials: Theory, multiscale methods and applications*. 2006: John Wiley & Sons Ltd.
11. Ma, J., H. Lu, B. Wang, R. Hornung, A. Wissink, and R. Komanduri, Multiscale simulation using generalized interpolation material point (GIMP) method and molecular dynamics (MD). *Computer Modeling in Engineering and Sciences*, 2006. **14**(2): p. 101–117.
12. Saether, E., V. Yamakov, and E. Glaessgen, An embeded statistical method for coupling molecular dynamics and finite element analyses. *International Journal for Numerical Methods in Engineering*, 2009. **78**: p. 1292–1319.
13. Reddy, J.N., *An introduction to the finite element method*, 4th ed. 2019: McGraw-Hill.
14. Needleman, A., A continuum model for void nucleation by inclusion debonding. *Journal of Applied Mechanics*, 1987. **54**: p. 525–531.
15. Needleman, A., An analysis of decohesion along an imperfect interface. *Journal of the Mechanics and Physics of Solids*, 1990. **38**(3): p. 289–324.
16. Tvergaard, V. and J.W. Hutchinson, The influence of plasticity on mixed mode interface toughness. *Journal of the Mechanics and Physics of Solids*, 1993. **41**: p. 1119–1135.
17. Tvergaard, V. and J.W. Hutchinson, Effect of T-stress on Mode I crack growth resistance in a ductile solid. *International Journal of Solids and Structures*, 1994. **31**: p. 823–833.
18. De Borst, R., Numerical aspects of cohesive-zone models. *Engineering Fracture Mechanics*, 2003. **70**: p. 1743–1757.
19. Knauss, W.G. and G.U. Losi, Crack propagation in a nonlinearly viscoelastic solid with relevance to adhesive bond failure. *Journal of Applied Mechanics*, 1993. **60**: p. 793–801.
20. Rybicki, E.F. and M.F. Kanninen, A finite element calculation of stress intensity factors by a modified crack closure integral. *Engineering Fracture Mechanics*, 1977. **9**(4): p. 931–938.
21. Krueger, R., The virtual crack closure technique: History, approach, and applications. NASA/TM-2002-211628. NASA Langley Research Center. 2002.
22. Krueger, R., Virtual crack closure technique: History, approach, and applications. *Applied Mechanics Reviews*, 2004. **57**(2): p. 109–143.
23. Zi, G. and T. Belytschko, New crack-tip elements for XFEM and applications to cohesive cracks. *International Journal for Numerical Methods in Engineering*, 2003. **57**(15): p. 2221–2240.
24. Griffith, A.A., VI. The phenomena of rupture and flow in solids. *Philosophical Transactions of the Royal Society A: Mathematical, Physical and Engineering Sciences*, 1921. **221**(582–593): p. 163–198. doi: 10.1098/rsta.1921.0006
25. Mumford, D. and J. Shah, Optimal approximations by piecewise smooth functions and associated variational problems. *Communications on Pure and Applied Mathematics*, 1989. **42**(5): p. 577–685. doi: 10.1002/cpa.3160420503
26. Ambrosio, L. and V.M. Tortorelli, Approximation of functional depending on jumps by elliptic functional via t-convergence. *Communications on Pure and Applied Mathematics*, 1990. **43**(8): p. 999–1036. doi: 10.1002/cpa.3160430805
27. Bourdin, B., G.A. Francfort, and J.-J. Marigo, Numerical experiments in revisited brittle fracture. *Journal of the Mechanics and Physics of Solids*, 2000. **48**(4); p. 797–826. doi: 10.1016/S0022-5096(99)00028-9
28. Francfort, G.A. and J.-J. Marigo, Revisiting brittle fracture as an energy minimization problem. *Journal of the Mechanics and Physics of Solids*, 1998. **46**(8): p. 1319–1342. doi: 10.1016/S0022-5096(98)00034-9

29. Amor, H., J.-J. Marigo, and C. Maurini, Regularized formulation of the variational brittle fracture with unilateral contact: Numerical experiments. *Journal of the Mechanics and Physics of Solids*, 2009. **57**(8): p. 1209–1229. doi: 10.1016/j.jmps.2009.04.011
30. Miehe, C., M. Hofacker, and F. Welschinger, A phase field model for rate-independent crack propagation: Robust algorithmic implementation based on operator splits. *Computer Methods in Applied Mechanics and Engineering*, 2010. **199**(45): p. 2765–2778. doi: 10.1016/j.cma.2010.04.011
31. Li, P. et al., A review on phase field models for fracture and fatigue. *Engineering Fracture Mechanics*, 2023. **289**: p. 109419. doi: 10.1016/j.engfracmech.2023.109419
32. Najafi Koopas, R., S. Rezaei, N. Rauter, R. Ostwald, and R. Lammering, Comparative analysis of phase-field and intrinsic cohesive zone models for fracture simulations in multiphase materials with interfaces: Investigation of the influence of the microstructure on the fracture properties. *Applied Sciences*, 2025. **15**(1): p. 160. doi: 10.3390/app15010160
33. Tran, H. and H.B. Chew, Cohesive zone interpretations of phase-field fracture models. *Journal of Applied Mechanics*, 2022. **89**(12): p. 1–21. doi: 10.1115/1.4055660
34. Cervera, M., G.B. Barbat, M. Chiumenti, and J.-Y. Wu, A comparative review of XFEM, mixed FEM and phase-field models for quasi-brittle cracking. *Archives of Computational Methods in Engineering*, 2022. **29**(2): p. 1009–1083. doi: 10.1007/s11831-021-09604-8

3 An Overview of Molecular Dynamics (MD) for Atomistic Simulations

3.1 BASIC CONCEPT OF MOLECULAR DYNAMICS (MD)

3.1.1 Newtonian Dynamics

MD involves the integration of Newton's equations of motion for each atom within an ensemble. A set of particles within an ensemble is initialized with positions and velocities. A set of interacting rules (interatomic potentials) specifies the forces each atom exert on its neighbors, based on relative position, electrical charge, etc. Next, the atoms are allowed to move over a short span of time (a single timestep), based on the interatomic forces. At the end of a timestep, the system consists of updated positions and velocities, and in turn, updated forces. The integration of Newton's equations of motion is performed numerically for all atoms in the ensemble. One of the integration algorithms used is the velocity-Verlet integration, which is shown in the schematic in Figure 3.1, where $\boldsymbol{r}^{(t)}$, $\boldsymbol{v}^{(t)}$, and $\boldsymbol{f}^{(t)}$ symbolize vectors containing atomic positions, velocities, and forces respectively, at time t, and m_i are atomic masses. The vectors are updated at each timestep (Δt) based on the equations shown in Figure 3.1. Subscript i refers to the ith individual atom.

For stable interatomic potential sets, the system of particles is expected to proceed toward energy equilibrium at the end of each timestep, obeying Newton's law.

3.1.2 Ensembles and Periodic Boundaries

The basic quantities in the discussion of statistical ensembles are the number of particles, volume, pressure, temperature, energy, enthalpy, and entropy. Some of the common ensembles used in MD are NVE (constant number of particles, volume, and energy), NVT (constant number of particles, volume, and temperature), NPT (constant number of particles, pressure, and temperature), and NPH (constant number of particles, pressure, and enthalpy).

The boundaries of the MD system may be periodic (e.g., when simulating an infinite body) or non-periodic (e.g., when accounting for finite-size effects/edge effects). Typically, to measure elastic properties of the bulk of a material such as a polymer matrix, the NPT ensemble with periodic boundaries is used, as described in Griebel and Hamaekers [1], which allows for Poisson's effect.

DOI: 10.1201/9781003440734-3

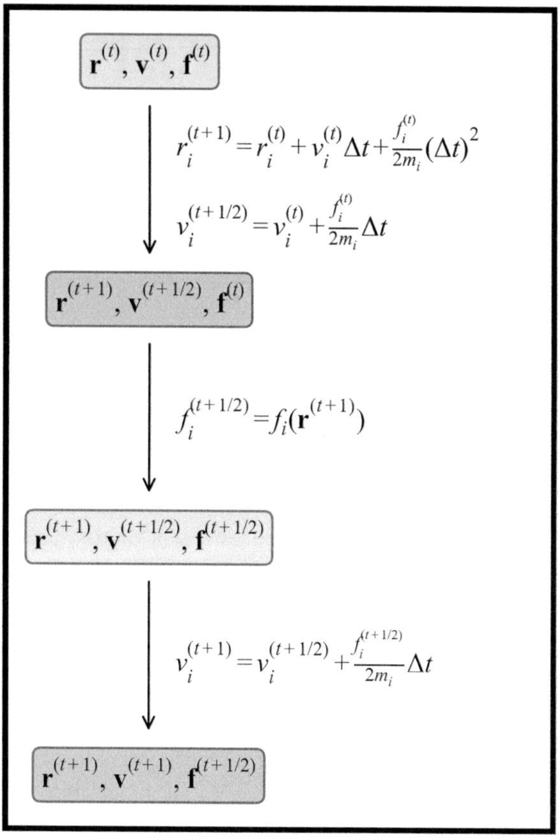

FIGURE 3.1 Flow chart of the velocity-Verlet integration algorithm.

3.1.3 Energy Equilibration

When any thermal or mechanical perturbation is applied to an MD ensemble that is already in equilibrium, sufficient time must be allowed for it to re-achieve energy equilibration, unlike in continuum solvers. Without proper equilibration, the results of the simulation will be unreliable. Figure 3.2 depicts energy equilibration carried out by the authors of this book for an amorphous polymer system that was being mechanically loaded in a stepwise manner at $T = 300$ K, based on a K-field approach, as will be described in Chapter 5. As shown in Figure 3.2, for this case, 80 ps of equilibration time was required for the potential energy to reach a plateau before a new mechanical load could be applied. This underscores the need to break up the loading in MD into multiple sub-steps to reduce the overall equilibration time.

3.1.4 Thermostats and Barostats

As described by the basic principles of statistical mechanics in Reif et al. [2], atoms do not have an intrinsic temperature. The temperature of an atom is defined by the

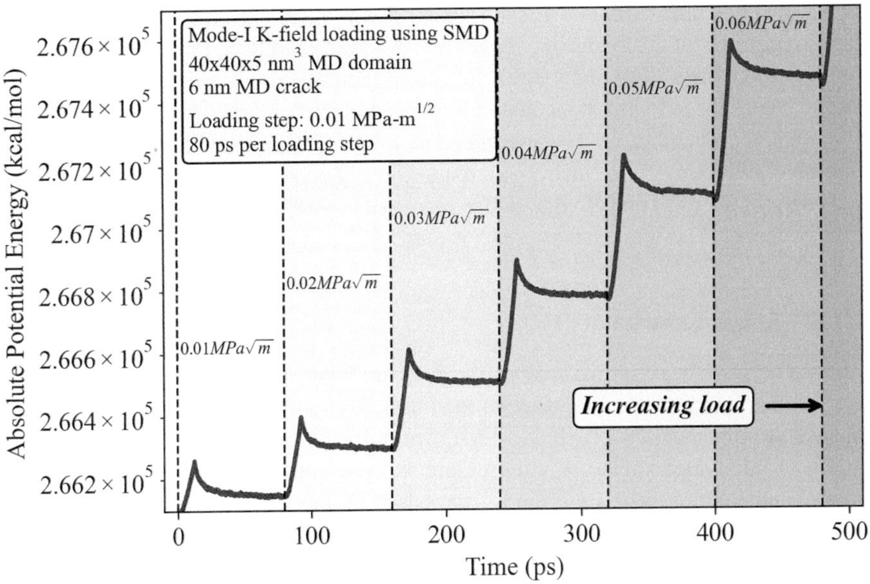

FIGURE 3.2 Potential energy equilibration over time for multiple loading steps during MD simulation of tensile fracture (K-test: refer to Chapter 5).

magnitude of the random vibration of the atom relative to its mean position, with the limiting condition that all thermally induced atomic motion ceases when temperature is lowered to absolute zero (0 K). For a three-dimensional MD ensemble of N atoms and a total kinetic energy, E_{kin}, the temperature of the ensemble is computed as follows:

$$T = \frac{2E_{\text{kin}}}{3Nk_B} \tag{3.1}$$

where k_B is the Boltzmann constant. It should be noted that the kinetic energy, E_{kin} in Equation (3.1) is due to the random thermal motion of the atoms within the ensemble relative to its center of mass, and not due to the motion of the center of mass itself.

Constant temperatures within an ensemble are maintained using thermostats. The Nosé–Hoover and Berendsen thermostats are common ones that perform velocity rescaling. The Andersen thermostat periodically reselects particle velocities from a Maxwell–Boltzmann distribution. The Langevin thermostat adds a friction term and noise to Newton's equations of motion, as described in Ke et al. [3].

Similar to thermostats, barostats are used to maintain a constant pressure. In MD, pressure involves a kinetic energy component, for which computing temperature is a prerequisite.

Another concept that is fundamental to MD and is the cornerstone of statistical mechanics, as described in Reif et al. [2], is entropy. Entropy signifies the number of states a system can be in that have equal probability. For instance, in an NVE ensemble, the vector of atomic positions at each timestep is a microstate with constant energy. Certain positions of the atoms may be more probable than others, for example, equilibrium positions in a crystalline solid. In glassy (amorphous) materials, however, the energy landscape is highly nonconvex; thus, there may be numerous microstates with nearly equal probability. In this case, the glassy material is said to have higher entropy than its equilibrium crystalline counterpart. Closely related quantities are the Gibbs free energy $(H - TS)$ and Helmholtz free energy $(U - TS)$, where H is enthalpy, U is internal energy, T is temperature, and S is entropy.

3.1.5 Stress Concepts

Stress is notionally a continuum concept, as given by the applied force divided by the area of cross section of a continuous medium. Applying the concept of stress to a discrete atomistic system, such as an MD ensemble, is nontrivial. An early work on atomistic calculations of stress and pressure was presented by Tsai [4]. Two methods of stress calculation were described—one based on the virial theorem and the other based on internal stress. The virial theorem is as follows:

$$PV = Nk_BT - \frac{1}{3}\left\langle \sum_{i,j<i}^{N} \boldsymbol{r}_{ij} \cdot \frac{\partial \Phi_{ij}}{\partial \boldsymbol{r}_{ij}} \right\rangle \tag{3.2}$$

where, P is the overall pressure in the ensemble, V is the volume, N is the number of atoms in the ensemble, T is the temperature of the system, k_B is Boltzmann's constant, $\boldsymbol{r}_{ij}$ is the position vector between atoms i and j, and Φ_{ij} is the interatomic potential between atoms i and j. The first term on the right side of Equation (3.2) is the kinetic energy term, and the second term corresponds to the "internal virial" due to interatomic forces. The virial formulation is well suited to measure the overall pressure for a system in equilibrium. The per-atom virial stress formulation can be found in Thompson et al. [5].

A detailed exploration of the techniques to calculate stress from atomistic simulations can be found in the work by, Admal and Tadmor [6]. The Irving–Kirkwood–Noll point-wise stress tensor, as described by Noll [7], lays the foundation for atomistic definitions of stress (Eqn. 3.63 in Admal and Tadmor [6]). Point-wise stress tensors need to be spatially averaged to arrive at continuum fields of stress. The stress definition proposed by Hardy [8], for the special case of pair potentials, is as follows:

$$\boldsymbol{\sigma}_v(\boldsymbol{x}, t) = \frac{1}{2}\sum_{\alpha \neq \beta} \frac{(\boldsymbol{x}_\alpha(t) - \boldsymbol{x}_\beta(t)) \otimes (\boldsymbol{x}_\alpha(t) - \boldsymbol{x}_\beta(t))}{\|\boldsymbol{x}_\alpha - \boldsymbol{x}_\beta\|} \mathcal{V}'_{\alpha\beta} b(\boldsymbol{x}; \boldsymbol{x}_\alpha, \boldsymbol{x}_\beta) \tag{3.3a}$$

$$\boldsymbol{\sigma}_k(\boldsymbol{x}, t) = -\sum_{\alpha} w(\boldsymbol{x}_\alpha(t) - \boldsymbol{x}) m_\alpha \boldsymbol{v}_\alpha^{\text{rel}}(t) \otimes \boldsymbol{v}_\alpha^{\text{rel}}(t) \tag{3.3b}$$

where $\boldsymbol{\sigma}_v$ and $\boldsymbol{\sigma}_k$ are the potential and kinetic energy components of the stress, respectively. $\boldsymbol{x}_\alpha$ and $\boldsymbol{x}_\beta$ denote the spatial positions of the interacting atoms, $\mathcal{V}'_{\alpha\beta}$ is the derivative of the pair potential between atoms α and β, b represents the bond function (Eqn. 4.16 in Admal and Tadmor [6]), v_α^{rel} is the relative velocity of atom α with respect to the continuum velocity (Eqn. 4.24 in Admal and Tadmor [6]) and $v_\alpha^{\text{rel}}(t) \otimes v_\alpha^{\text{rel}}(t)$ implies the dyadic (tensor) product of the atomic velocities.

3.2 TYPES OF ATOMISTIC POTENTIALS

Atomistic interactions range from the simplest pair potentials like the Lennard-Jones potential to highly complex reactive potentials, like ReaxFF. The choice of potential depends on the problem to be solved and the type of phenomena being simulated. Covering all currently available potentials is not possible in the interest of space, and therefore, this chapter focuses on atomistic potentials which have application in fracture and fatigue of materials. An important aspect related to modeling fracture using the atomistic force field is the ability to model elastic and inelastic deformation, to be ultimately followed by bond breakage when fracture initiates.

3.2.1 Non-bonded Interactions

The basic Lennard-Jones (L-J) potential, along with a Coulomb interaction term, is shown below, which allows for "non-bonded" London forces interactions as well as bond breakage.

$$E = 4\epsilon\left[\left(\frac{\sigma}{r}\right)^{12} - \left(\frac{\sigma}{r}\right)^{6}\right] + \frac{Cq_iq_j}{\epsilon r} \tag{3.4}$$

where ϵ is the dielectric constant, σ is the zero-crossing distance, C is an energy conversion constant, and q_i and q_j are electrical charges on the interacting atoms. The L-J potential does allow for the stretching and final separation of non-bonded interactions between bi-material interfaces. It should be noted that not all potentials allow for bond breakage. For example, the widely used OPLS potential, as described by Jorgensen and Tirado-Rives [9], due to its harmonic construct, is not capable of modeling bond breakage and hence cannot be used for accurate fracture simulations.

3.2.2 Potentials for Metals

In the work by Andric and Curtin [10], fracture in single-crystal nickel is modeled using the Ni EAM (Embedded Atom Method) potential, as described in Mishin et al. [11]. The functional form of the EAM potential energy for atom i is given by

$$E_i = F_\alpha\left(\sum_{j\neq i}\rho_\beta(r_{ij})\right) + \frac{1}{2}\sum_{j\neq i}\phi_{\alpha\beta}(r_{ij}) \tag{3.5}$$

where F is an embedding energy function, ρ is the electron density, and ϕ is the pair potential term. The summation of all atomic neighbors within a cutoff radius makes the EAM potential a multi-body force field, as described in Bitzek et al. [12].

It should be noted that the majority of work related to fracture has been conducted using the EAM potential; however, these results suffer from quantitative inaccuracies compared with experimental results. The authors in Bitzek et al. [12] suggested the importance of the development of material-specific, chemically accurate atomistic potentials, using machine learning or bond-order based potentials. More fundamental quantum mechanics-based approaches, such as DFT, potentially offer highly accurate simulations of fracture. However, they are vastly limited in terms of computational efficiency and in their ability to model the required number of atoms to approach length and time scales necessary for simulating practical applications. More complex studies on metal fracture using atomistic simulations such as the effect of hydrogen embrittlement, as described in Tehranchi and Curtin [13], require the use of specific adaptations to the EAM potential.

3.2.3 Potentials for Non-metals

Andric and Curtin [10] used the modified EAM (MEAM) potential, as described in Cui et al. [14], to simulate fracture in a Li–Si system. MEAM is an extension of the EAM potential with added angular forces. Aside from metals, it can be used to model covalent interactions in materials based on carbon and silicon, as described in Ko and Lee [15], etc.

In Dontsova and Ballarini [16], a set of nine atomistic potentials, including the three-body potentials (as described in Tersoff and Kumagai et al. [17, 18] and Stillinger and Weber, Holland and Marder [19, 20]) and the MEAM potential, were compared in the simulation of fracture in silicon.

For fracture simulations involving amorphous carbon, as described in Khosrownejad et al. [21] used a variant of the Tersoff potential, as described in Pastewka et al. [22].

Potentials such as ReaxFF, as described in Senftle et al. [23], are unique in the sense that they allow for bond breakage as well as chemical reactions to occur within the MD ensemble. They are based on bond order, which is determined as a function of interatomic distance. While ReaxFF is very computationally intensive, it provides more reliable fracture results. In Nishino et al. [24], ReaxFF was used to study the interfacial fracture of alumina/epoxy in the presence of water molecules.

ReaxFF consists of a sum of various energy terms, which includes bonds, three-body valence angles, four-body torsion angles, Van der Waals and Coulomb forces, etc. as given below in Equation (3.6).

$$E_{\text{ReaxFF}} = E_{\text{bond}} + E_{\text{angle}} + E_{\text{over}} + E_{\text{tors}} + E_{\text{vdWaals}} + E_{\text{Coulomb}} + E_{\text{specific}} \quad (3.6)$$

ReaxFF was also used to simulate fracture in silica glass, as described in Park and Kirane, Rimsza et al. [25, 26], and the simulation results were in reasonable agreement with available experimental results.

In terms of non-metallic systems, such as polymers, MD-based fracture simulations have been somewhat limited, primarily due to large process zone sizes (~12,000 nm) near the crack tip.

For systems involving macromolecules such as polymers, it is often computationally efficient to use fixed bond potentials. The OPLS potential, as described in

Jorgensen and Tirado-Rives [9], and the COMPASS force field, as described in Sun et al. [27], are some examples of fixed-bond potentials. Bonds, angles, dihedrals, and impropers need to be specified by the user prior to the simulation, and they remain unchanged over the course of the simulation. Often, they are defined as harmonic potentials such as:

$$E_{\text{harmonic}} = K(r - r_0)^2 \tag{3.7}$$

where K is a spring constant, and r_0 is the equilibrium bond length. As mentioned earlier in this chapter, harmonic potentials, however, are accurate only for small strains and cannot simulate bond breakage accurately, and therefore cannot be used for fracture simulations. Using a Morse potential, instead of a harmonic potential to model fixed bonds, is one way to simulate bond breakage, which is a requirement for fracture:

$$E_{\text{morse}} = D\left[1 - e^{-\alpha(r - r_0)}\right]^2 \tag{3.8}$$

where E_{morse} is the Morse potential, α is a stiffness parameter, and D quantifies the potential well depth. In recent work, the INTERFACE force field, as described in Heinz et al. [28], has been augmented with the Morse potential to enable reactivity, as described in Winetrout et al. [29].

3.3 SPECIAL CONSIDERATIONS FOR MECHANICAL TESTING

MD simulations have been used to simulate a wide variety of chemical and physical processes in the literature including hygrothermal infiltration, as described in Hassanali and Singer [30], corrosion and oxidation, as described in Shen et al. [31], particle impact, as described in Ogawa [32], pyrolysis, as described in Zheng et al. [33], polymer crosslinking, as described in Vashisth et al. [34], among others. The application of MD to performing simulated mechanical tests on solid materials requires some special considerations. Length-scale and time-scale aspects such as strain rate, domain size effect, boundary conditions, etc., come into play and must be taken into account.

3.3.1 Strain Rate Effects in MD Simulations

MD simulations typically have timesteps in the order of femtoseconds (1 fs = 10^{-15} s). For reactive potentials such as ReaxFF, the timestep can be as low as 0.1 fs. Therefore, the overall simulation timescales that can be achieved using MD are quite small relative to actual experiments. For example, with a system of approximately 100,000 atoms modeled with the OPLS potential, the computational performance on 64 parallel processors is approximately 7 ns of analysis time per day. Experimentally, high strain rates are usually ~10^3/s, as described in Nicholas, Davis [35, 36]. However, ultra-high strain rates of ~10^6/s have been achieved using Taylor impact tests, as described in Piao et al. [37]. But, in MD simulations, computationally feasible strain rates are typically higher than 10^8 s^{-1} due to the maximum allowable time-step constraint. Therefore, addressing strain-rate effects resulting from extremely high strain rates used in MD simulations is an important aspect to consider for mechanical test simulations using MD, especially when comparing simulation results with experimental data for a strain-rate-sensitive material.

In the literature, MD studies have been conducted in various materials at strain rates typically varying between 5×10^7 and 5×10^{10}/s. In metallic materials, various MD simulation strain rates have produced varying deformation mechanisms, as described in Koh et al. [38]. For epoxy polymers, Li and Strachan [39] observed insensitivity of the Young's modulus on strain rate. However, the yield stress showed strong dependence on strain rate, with higher strain rates resulting in higher yield stress (~20% increase). Materials such as graphene exhibit lower sensitivity of tensile strength on strain rate (~4% increase), as described in Chen et al. [40]. For amorphous polyethylene, Sahputra and Echtermeyer [41] found significant variation in the Young's modulus (>50% in some cases) with varying strain rates. Therefore, assessing strain rate effects in MD simulations is a key aspect of making reliable predictions of mechanical properties, and of making apples-to-apples comparison with experimental results.

Methods such as power-law least squares fitting have been used in the literature, as described in Odegard et al. [42], to extrapolate and compare MD predicted Young's moduli at high strain rates with available experimental data at much lower strain rates. Other more complex fitting functions have been used to extrapolate MD predicted strength data to experimentally relevant strain rates, as described in Chowdhury and Gillespie [43]. The time-temperature superposition approach, based on "master curves," has also been used with limited success in the literature, as described in Sahputra and Echtermeyer [41], to extrapolate high-strain-rate MD data to lower strain rates. However, no physics-based general methodology has been proposed for addressing the strain rate issue.

3.3.2 Size Effects in MD Simulations

The effect of MD domain size is another critical aspect of MD simulation that must be properly resolved. Typically, the prediction of elastic and yield properties using MD simulations makes use of periodic unit cells or representative volume elements (RVE). The size of the RVE is critical to ensuring that sufficient spatial variation in the material nanostructure is included. A thorough analysis of RVE size effect on the Young's modulus predicted in epoxy resins simulated using MD was conducted by Wan et al. [44]. Significant statistical variations in modulus were found to occur when the RVE size was below a certain length, based on multiple MD realizations of the RVE. The effect of specimen size, crack length, etc., are other important considerations when dealing with finite geometries with free edges such as nanowires, as described in Guo et al. [45], who reported significant increase in Young's modulus, yield stress, and tensile strength with increasing cross-sectional area of diamond nanowires.

Brittle fracture problems are inherently size dependent when the small-scale yielding (SSY) criterion is not fulfilled, as will be discussed in detail in Chapter 5.

Park and Kirane [25] showed how various crack lengths and specimen widths in single-edge-notched tension (SENT) tests on silica glass exhibited varying fracture behavior, with transition from brittle fracture to ductile rupture below a critical crack length. Similar fracture tests on amorphous carbon materials with size dependency were studied by Lu et al. [46]. Therefore, careful consideration of MD domain size based on the specific material being studied is imperative, or the predicted values will be unrealistic.

3.3.3 Non-periodic Boundaries and Steered MD

For fracture simulations, specifically, performing fully periodic RVE-based simulations is usually not feasible. Instead, the MD domain needs to have some free edges, which can be intentionally made stress free (e.g., nanowires) or theoretically infinite (e.g., while performing a K-test approach, as will be described in Chapter 5). The presence of free edges introduces additional challenges not present when using periodic boundary conditions. For example, polymer chains need to be appropriately terminated at the crack surface, without leaving hanging free radicals. The lack of connectivity at the free edges can result in reduced material stiffness near the free edges, as explained in Pfaller et al. [47]. Techniques such as dissipative particle dynamics (DPD), as described in Espanol and Warren [48], or Langevin thermostatting, as described in Schneider and Stoll, Dunweg and Paul [49, 50], are usually recommended in boundary regions of MD domains, which can help to emulate an infinite body.

Fracture simulations with free-edged MD domains typically need to be loaded using certain boundary conditions applied to boundary regions of the MD domain. One approach is to not allow time integration of the boundary atoms, thereby effectively "freezing them." Displacements can directly be applied to these frozen atoms as was done in some studies in the literature, as described in Andric and Curtin, and Rimsza et al. [10, 26]. Another approach, which will be explained in detail in later chapters, is to use steered MD (SMD), as described in Izrailev et al. [51]. SMD is a robust technique based on spring forces being used to gradually apply conformational changes to MD atomistic systems, instead of instantaneously applied deformation that could cause instabilities. The degree of coupling can be controlled through the value of the spring stiffness. A high value of the spring stiffness would result in a direct one-to-one coupling with a high spring force, whereas a low value will result in a weaker coupling with a low spring force. A similar concept was used by Pfaller et al. [47, 52], and later extended by Aditya et al. [53] to couple MD domains with FEM domains.

3.3.4 Computation of Virial Stress

As alluded to in Section 3.1.5, the virial stress, as described in Tsai [4], is an essential part of atomistic simulations, involving the extraction of continuum stress measures from discrete atomistic data such as forces and velocities. It can be expressed as follows,

$$\tau_{ij} = \frac{1}{\Omega} \sum_{k \in \Omega} \left[-m^{(k)} \left(u_i^{(k)} - \overline{u}_i \right) \left(u_j^{(k)} - \overline{u}_j \right) + \frac{1}{2} \sum_{l \in \Omega} \left(x_i^{(l)} - x_i^{(k)} \right) f_j^{(kl)} \right] \tag{3.9}$$

where τ_{ij} are the scalar components of the virial stress tensor, k and l are atoms lying within the domain Ω, $m^{(k)}$ is the mass of atom k, $u_i^{(k)}$ are velocity components of the kth atom, $\overline{u}_i$ are velocity components of the average of all atoms in the volume Ω, $x_i^{(l)}$ are atomic position components of the lth atom, and $f_j^{(kl)}$ are components of the force between atoms k and l. For a three-dimensional MD system, i and j span from one to three. Typically, molecular dynamics codes such as Large-scale Atomic/Molecular Massively Parallel Simulator (LAMMPS) output virial stress per atom, as described in Thompson et al. [5], in energy units (the term within square brackets in Equation (3.9)). Defining the volume of individual atoms is not straightforward, although techniques such as Voronoi tessellations can be useful in certain situations. A common approach is to consider stress computation boxes around evaluation points and sum up the values of the virial stress per atom within that box, followed by dividing by the volume of the box, Ω, as indicated by the $\frac{1}{\Omega}$ term in Equation (3.9). This technique is usually combined with temporal averaging of the per-atom virial stress over certain time periods to smooth out thermal oscillations (at finite temperatures) about an equilibrium value. In order for the computed virial stress to be box-size independent, the approach must involve a box-size convergence analysis to ensure the computation box is sufficiently large to homogenize the discrete atomic stresses, as described by Roy and Roy [54]. A more comprehensive literature study of the available techniques to estimate continuum stress tensors from MD can be found in Admal and Tadmor [6].

3.4 SUMMARY

An overview of molecular dynamics simulation was presented in this chapter, right from essential concepts of dynamics equations, statistical ensembles, thermostatting/barostatting, and energy equilibration to practical considerations required to apply MD simulations in fracture modeling. A brief review of various interatomic potentials was provided, including potentials for metallic and non-metallic materials. The importance of modeling non-bonded interactions using the Lennard-Jones potential was underscored for modeling failure at interfaces that are not covalently bonded. Because MD ensembles differ from traditional continuum systems due to their thermal motion about their mean atomic positions, the requirement that thermal equilibrium must be attained at each load step is essential for MD simulations to produce accurate results was highlighted. The concept of virial stress was introduced, which will be used in later chapters to compute stresses near the crack tip (Chapter 5) as well as for computing the atomistic J-integral (Chapter 4) as a fracture metric. Finally, practical issues that need to be addressed when performing mechanical loading simulations were elucidated, including virial stress computation, strain rate effects, domain size effects, temporal averaging, periodicity, and free edge effects. The foundations laid in this chapter will serve as the basis for the complex, practical simulations to be discussed in the remainder of this book.

REFERENCES

1. Griebel, M. and J. Hamaekers, Molecular dynamics simulations of the elastic moduli of polymer–carbon nanotube composites. *Computer Methods in Applied Mechanics and Engineering*, 2004. **193**(17–20): p. 1773–1788.
2. Reif, F., *Fundamentals of statistical and thermal physics*. 2009: Waveland Press.
3. Ke, Q., et al., Effects of thermostats/barostats on physical properties of liquids by molecular dynamics simulations. *Journal of Molecular Liquids*, 2022. **365**: p. 120116.
4. Tsai, D., The virial theorem and stress calculation in molecular dynamics. *The Journal of Chemical Physics*, 1979. **70**(3): p. 1375–1382.
5. Thompson, A.P., S.J. Plimpton, and W. Mattson, General formulation of pressure and stress tensor for arbitrary many-body interaction potentials under periodic boundary conditions. *The Journal of Chemical Physics*, 2009. **131**(15): p. 154107.
6. Admal, N.C. and E.B. Tadmor, A unified interpretation of stress in molecular systems. *Journal of Elasticity*, 2010. **100**: p. 63–143.
7. Noll, W., Die Herleitung der Grundgleichungen der Thermomechanik der Kontinua aus der statistischen Mechanik. *Journal of Rational Mechanics and Analysis*, 1955. **4**: p. 627–646.
8. Hardy, R.J., Formulas for determining local properties in molecular-dynamics simulations: Shock waves. *The Journal of Chemical Physics*, 1982. **76**(1): p. 622–628.
9. Jorgensen, W.L. and J. Tirado-Rives, Potential energy functions for atomic-level simulations of water and organic and biomolecular systems. *Proceedings of the National Academy of Sciences*, 2005. **102**(19): p. 6665–6670.
10. Andric, P. and W.A. Curtin, Atomistic modeling of fracture. *Modelling and Simulation in Materials Science and Engineering*, 2018. **27**(1): p. 013001.
11. Mishin, Y., et al., Interatomic potentials for monoatomic metals from experimental data and ab initio calculations. *Physical Review B*, 1999. **59**(5): p. 3393.
12. Bitzek, E., J.R. Kermode, and P. Gumbsch, Atomistic aspects of fracture. *International Journal of Fracture*, 2015. **191**: p. 13–30.
13. Tehranchi, A. and W.A. Curtin, The role of atomistic simulations in probing hydrogen effects on plasticity and embrittlement in metals. *Engineering Fracture Mechanics*, 2019. **216**: p. 106502.
14. Cui, Z., et al., A second nearest-neighbor embedded atom method interatomic potential for Li–Si alloys. *Journal of Power Sources*, 2012. **207**: p. 150–159.
15. Ko, W.-S. and B.-J. Lee, Origin of unrealistic blunting during atomistic fracture simulations based on MEAM potentials. *Philosophical Magazine*, 2014. **94**(16): p. 1745–1753.
16. Dontsova, E. and R. Ballarini, Atomistic modeling of the fracture toughness of silicon and silicon-silicon interfaces. *International Journal of Fracture*, 2017. **207**: p. 99–122.
17. Tersoff, J., Empirical interatomic potential for silicon with improved elastic properties. *Physical Review B*, 1988. **38**(14): p. 9902.
18. Kumagai, T., et al., Development of bond-order potentials that can reproduce the elastic constants and melting point of silicon for classical molecular dynamics simulation. *Computational Materials Science*, 2007. **39**(2): p. 457–464.
19. Stillinger, F.H. and T.A. Weber, Computer simulation of local order in condensed phases of silicon. *Physical Review B*, 1985. **31**(8): p. 5262.
20. Holland, D. and M. Marder, Ideal brittle fracture of silicon studied with molecular dynamics. *Physical Review Letters*, 1998. **80**(4): p. 746.
21. Khosrownejad, S.M., J.R. Kermode, and L. Pastewka, Quantitative prediction of the fracture toughness of amorphous carbon from atomic-scale simulations. *Physical Review Materials*, 2021. **5**(2): p. 023602.
22. Pastewka, L., et al., Screened empirical bond-order potentials for Si–C. *Physical Review B*, 2013. **87**(20): p. 205410.

23. Senftle, T.P., et al., The ReaxFF reactive force-field: Development, applications and future directions. *npj Computational Materials*, 2016. **2**(1): p. 1–14.
24. Nishino, H., et al., Fracture behavior of an alumina/epoxy resin interface and effect of water molecules by using molecular dynamics with reaction force field (ReaxFF). *International Journal of Adhesion and Adhesives*, 2022. **119**: p. 103251.
25. Park, J. and K. Kirane, Transitional flaw size sensitivity of amorphous silica nanostructures analyzed by ReaxFF/SiO based molecular dynamics. *Journal of Applied Physics*, 2021. **129**(17): p. 175103.
26. Rimsza, J.M., R.E. Jones, and L.J. Criscenti, Crack propagation in silica from reactive classical molecular dynamics simulations. *Journal of the American Ceramic Society*, 2018. **101**(4): p. 1488–1499.
27. Sun, H., P. Ren, and J. Fried, The COMPASS force field: Parameterization and validation for phosphazenes. *Computational and Theoretical Polymer Science*, 1998. **8**(1–2): p. 229–246.
28. Heinz, H., et al., Thermodynamically consistent force fields for the assembly of inorganic, organic, and biological nanostructures: the INTERFACE force field. *Langmuir*, 2013. **29**(6): p. 1754–1765.
29. Winetrout, J.J., et al., Implementing reactivity in molecular dynamics simulations with the interface force field (IFF-r) and other harmonic force fields. arXiv preprint arXiv: 2107.14418, 2021.
30. Hassanali, A.A. and S.J. Singer, Model for the water−amorphous silica interface: The undissociated surface. *The Journal of Physical Chemistry B*, 2007. **111**(38): p. 11181–11193.
31. Shen, F., et al., Corrosion and oxidation on iron surfaces in chloride contaminated electrolytes: Insights from ReaxFF molecular dynamic simulations. *Journal of Materials Research and Technology*, 2024. **29**: p. 1305–1312.
32. Ogawa, H., Molecular dynamics simulation on the single particle impacts in the aerosol deposition process. *Materials Transactions*, 2005. **46**(6): p. 1235–1239.
33. Zheng, M., et al., Initial chemical reaction simulation of coal pyrolysis via ReaxFF molecular dynamics. *Energy & Fuels*, 2013. **27**(6): p. 2942–2951.
34. Vashisth, A., et al., Accelerated ReaxFF simulations for describing the reactive cross-linking of polymers. *The Journal of Physical Chemistry A*, 2018. **122**(32): p. 6633–6642.
35. Nicholas, T., Tensile testing of materials at high rates of strain: An experimental technique is developed for testing materials at strain rates up to 103 s− 1 in tension using a modification of the split Hopkinson bar or Kolsky apparatus. *Experimental Mechanics*, 1981. **21**(5): p. 177–185.
36. Davis, J.R., *Tensile testing*. 2004: ASM International.
37. Piao, M., et al., Characterization of flow stress at ultra-high strain rates by proper extrapolation with Taylor impact tests. *International Journal of Impact Engineering*, 2016. **91**: p. 142–157.
38. Koh, S., et al., Molecular dynamics simulation of a solid platinum nanowire under uniaxial tensile strain: Temperature and strain-rate effects. *Physical Review B—Condensed Matter and Materials Physics*, 2005. **72**(8): p. 085414.
39. Li, C. and A. Strachan, Molecular dynamics predictions of thermal and mechanical properties of thermoset polymer EPON862/DETDA. *Polymer*, 2011. **52**(13): p. 2920–2928.
40. Chen, M., et al., Effects of grain size, temperature and strain rate on the mechanical properties of polycrystalline graphene – A molecular dynamics study. *Carbon*, 2015. **85**: p. 135–146.
41. Sahputra, I.H. and A.T. Echtermeyer, Effects of temperature and strain rate on the deformation of amorphous polyethylene: A comparison between molecular dynamics simulations and experimental results. *Modelling and Simulation in Materials Science and Engineering*, 2013. **21**(6): p. 065016.

42. Odegard, G.M., et al., Predicting mechanical response of crosslinked epoxy using ReaxFF. *Chemical Physics Letters*, 2014. **591**: p. 175–178.
43. Chowdhury, S.C. and J.W. Gillespie Jr, Strain-rate dependent mode I cohesive traction laws for glass fiber-epoxy interphase using molecular dynamics simulations. *Composites Part B: Engineering*, 2022. **237**: p. 109877.
44. Wan, S., R.C. Sinclair, and P.V. Coveney, Uncertainty quantification in classical molecular dynamics. *Philosophical Transactions of the Royal Society A*, 2021. **379**(2197): p. 20200082.
45. Guo, J., et al., Molecular dynamics study on diamond nanowires mechanical properties: Strain rate, temperature and size dependent effects. *Diamond and Related Materials*, 2011. **20**(4): p. 551–555.
46. Lu, Q., et al., Nanoscale fracture of tetrahedral amorphous carbon by molecular dynamics: Flaw size insensitivity. *Physical Review B—Condensed Matter and Materials Physics*, 2008. **77**(1): p. 014109.
47. Pfaller, S., A. Kergaßner, and P. Steinmann, Optimisation of the capriccio method to couple particle-and continuum-based simulations of polymers. *Multiscale Science and Engineering*, 2019. **1**: p. 318–333.
48. Espanol, P. and P. Warren, Statistical mechanics of dissipative particle dynamics. *Europhysics Letters*, 1995. **30**(4): p. 191.
49. Schneider, T. and E. Stoll, Molecular-dynamics study of a three-dimensional one-component model for distortive phase transitions. *Physical Review B*, 1978. **17**(3): p. 1302.
50. Dünweg, B. and W. Paul, Brownian dynamics simulations without Gaussian random numbers. *International Journal of Modern Physics C*, 1991. **2**(03): p. 817–827.
51. Izrailev, S., et al. Steered molecular dynamics. in *Computational Molecular Dynamics: Challenges, Methods, Ideas: Proceedings of the 2nd International Symposium on Algorithms for Macromolecular Modelling*, Berlin, May 21–24, 1997. 1999. Springer.
52. Pfaller, S., et al., An Arlequin-based method to couple molecular dynamics and finite element simulations of amorphous polymers and nanocomposites. *Computer Methods in Applied Mechanics and Engineering*, 2013. **260**: p. 109–129.
53. Aditya, S.S., T. Sohail, and S. Roy, An atomistic-continuum concurrent statistical coupling technique for amorphous materials using anchor points. *Modelling and Simulation in Materials Science and Engineering*, 2023. **31**(7): p. 075012.
54. Roy, S. and A. Roy, A computational investigation of length-scale effects in the fracture behaviour of a graphene sheet using the atomistic J-integral. *Engineering Fracture Mechanics*, 2019. **207**: p. 165–180.

4 Development of the Atomistic *J*-integral
Energetic and Entropic Effects

4.1 INTRODUCTION

Fracture can be defined as the separation of a structure into two or more pieces when subjected to mechanical loading. It is associated with the development of a displacement discontinuity in the structure, followed by crack growth. When the direction of loading is perpendicular to the direction of crack growth, it is called Mode-I fracture, as discussed in Chapter 1.

Fracture can be characterized through parameters such as strain energy release rate (G), stress intensity factor (K), and the J-integral (J). The critical values of these parameters represent the fracture strength of a material. Once the critical value is reached, crack initiation occurs in the material. While G and K are widely used to characterize fracture driving force, they are applicable only in the linear elastic range of the stress-strain response of the material. On the other hand, the J-integral is identical to G in the linear range but has the advantage of being applicable in the nonlinear range as well as being contour path independent. Also, the critical value of J-integral at crack initiation is directly related to fracture toughness of the material, while the value of J as a function of crack length (a) during crack growth provides the resistance of the material to crack growth, i.e., the R-curve. Therefore, the J-integral could be used as a fracture criterion at the nanoscale, provided some modifications are implemented to the conventional definition of J.

4.2 ATOMISTIC *J*-INTEGRAL EVALUATION METHODOLOGY

The J-integral can be used to quantify the crack driving force available from thermomechanical loading as well as due to material inhomogeneities. The concept of the J-integral was developed independently by Rice [1] and by Eshelby [2]. In conventional macro-scale fracture mechanics, the total J-integral vector (J_T) is defined as the volume integral of the divergence of the Eshelby energy-momentum tensor [2] and is shown in Equation (4.1) after it has been reduced to a surface integral through the application of the divergence theorem.

$$\boldsymbol{J}_T = \int_{\partial\Omega} \langle \underline{S} \rangle \boldsymbol{N} dA = \int_{\partial\Omega} \left(\langle \Psi \rangle \boldsymbol{N} - \langle \underline{H}^T \underline{P} \rangle \boldsymbol{N} \right) dA = \boldsymbol{J}_U - \boldsymbol{J}_\eta \tag{4.1}$$

DOI: 10.1201/9781003440734-4

In Equation (4.1), $\underline{S}$ is the Eshelby energy-momentum tensor, Ψ is the Helmholtz free energy density, $\underline{H}$ is the displacement gradient tensor, $\underline{P}$ is the first Piola-Kirchhoff stress tensor, N is the outward normal to the surface $\partial\Omega$ along which contour the J-integral is being evaluated, J_U is the contribution to the J-integral due to internal energy (U), and J_η is the entropic contribution where η is the entropy (J_U and J_η will be discussed in detail in the following sections). Here it is assumed that the ensemble average $\langle * \rangle$ is approximated by the time average of the quantity over a sufficiently long period of time. The critical value of scalar component J_k at crack initiation is related to the fracture toughness of the material, where the subscript k denotes Cartesian components of the J vector $(k = 1, 2, 3)$ in three-dimensions and these components are related to but are not the same as the three primary fracture modes. As discussed in Chapter 1, $J_1 = J^{\mathrm{I}} + J^{\mathrm{II}} + J^{\mathrm{III}}$ and $J_2 = -2\sqrt{(J^{\mathrm{I}}J^{\mathrm{II}})}$, where the subscripts imply Cartesian components of the J-vector, and the superscripts imply standard fracture modes, as described in Rigby and Aliabadi [3]. In the absence of Mode-III, i.e., for purely in-plane deformation, the above equations can be solved to obtain the individual fracture modes from the Cartesian components of the J vector as given by, $J^{\mathrm{I}} = \frac{J_1 \pm \sqrt{J_1^2 - J_2^2}}{2}$, and $J^{\mathrm{II}} = J_1 - J^{\mathrm{I}}$. Therefore, the Cartesian components of the J-integral can be used as a suitable metric for estimating the crack driving force as well as the fracture toughness of a material as the crack begins to initiate, for Mode-I, Mode-II, and for mixed-mode fracture processes undergoing proportional loading. However, for the conventional macroscale definition of the J-integral to be valid at the nanoscale in terms of the continuum stress and displacement fields (and their spatial derivatives) requires the construction of local continuum fields from discrete atomistic data, and using these data in the conventional contour integral expression for J, as given by Equation (4.1) based on work by Jones and Zimmerman [4, 5]. One such methodology was proposed by Hardy [6], which allows for the local averaging necessary to obtain the definition of free energy, deformation gradient, and Piola-Kirchhoff stress as fields and gradients of fields, and not just as total system averages. The formulae used for evaluating each term on the RHS in Equation (4.1) from MD simulations are given below.

The energetic component of total J (J_U) is computed using the formula,

$$\boldsymbol{J}_U = \int_{\partial\Omega} \left(\langle U \rangle \boldsymbol{N} - \langle \underline{H}^{\mathrm{T}} \underline{P} \rangle \boldsymbol{N} \right) dA \tag{4.2}$$

In order to facilitate the J-integral computation using discrete atomistic data, localization boxes, as shown in Figure 4.1, are constructed along the integral contour, and the position of an atom with respect to the box determines the value of localization function. The localization function has to satisfy the following conditions: $\psi > 0$ and $\int_{\Omega} \psi dV = 1$. The localization function used in this chapter is given by Equation (4.3).

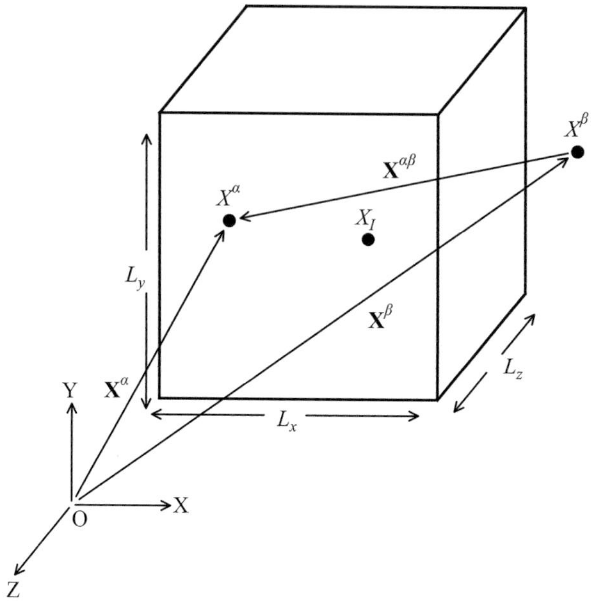

FIGURE 4.1 Illustration of a localization box with one atom inside and another outside the box.

$$\zeta(\boldsymbol{X}-\boldsymbol{X}^{\alpha}) = \begin{cases} \dfrac{1}{L_x L_y L_z} & \text{if } \begin{array}{l} X_I - \dfrac{L_x}{2} \leq X^{\alpha} \leq X_I + \dfrac{L_x}{2} \\ Y_I - \dfrac{L_y}{2} \leq Y^{\alpha} \leq Y_I + \dfrac{L_y}{2} \\ Z_I - \dfrac{L_z}{2} \leq Z^{\alpha} \leq Z_I + \dfrac{L_z}{2} \end{array} \\ 0 & \end{cases} \tag{4.3}$$

where L_x, L_y and L_z are the dimensions of the localization box and X_I, Y_I, and Z_I are the coordinates of the centroid, as illustrated in Figure 4.1. The terms required for the computation of J_U can now be defined as,

$$U(\boldsymbol{X},t) = \sum_{I=1}^{K}\sum_{\alpha=1}^{M} N_I(\boldsymbol{X})\zeta_{I\alpha}(\boldsymbol{X}^{\alpha}-\boldsymbol{X}_I)\phi^{\alpha}(t) - U_R(\boldsymbol{X}) \tag{4.4}$$

$$\underline{H}(\boldsymbol{X},t) = \nabla_{\boldsymbol{x}}\boldsymbol{u}(\boldsymbol{X},t) = \sum_{I=1}^{K}\sum_{\alpha=1}^{M} \nabla_{\boldsymbol{x}} N_I(\boldsymbol{X})\zeta_{I\alpha}(\boldsymbol{X}^{\alpha}-\boldsymbol{X}_I)\boldsymbol{u}^{\alpha}(t) \tag{4.5}$$

$$\underline{P}(\boldsymbol{X},t) = -\frac{1}{2}\sum_{\alpha=1}^{M}\sum_{\beta\neq\alpha}^{M} \boldsymbol{f}^{\alpha\beta} \bigotimes \boldsymbol{X}^{\alpha\beta} B^{\alpha\beta}(\boldsymbol{X}) \tag{4.6}$$

In Equation (4.4), ϕ^{α} is the atomistic potential of atom α, $U_R(\boldsymbol{X})$ is the reference potential energy density of atom α at 0 K, ζ is the Hardy localization function, M is the number of atoms in the localization box, and $N_I(\boldsymbol{X})$ are interpolation functions, with $I = 1, K$, where K is the total number of nodes in the localization box. In Equation (4.5), $\boldsymbol{u}(\boldsymbol{X}, t)$ is the displacement vector. In Equation (4.6), $\boldsymbol{f}^{\alpha\beta}$ is the vector representing the force between atoms α and β, $\boldsymbol{X}^{\alpha\beta}$ is the vector representing the difference in their positions, and $B^{\alpha\beta}$ is the bond function defined by Hardy [6]. Detailed derivation of Equations (4.4)–(4.6) follows.

4.2.1 Evaluation of Energy Density Field ($\boldsymbol{U}$):

$$U(\boldsymbol{X}, t) = \sum_{\alpha=1}^{M} \left(\phi^{\alpha}(t) - \phi_X^{\alpha}\right)\zeta(\boldsymbol{X} - \boldsymbol{X}^{\alpha}) \tag{4.7}$$

$$U(\boldsymbol{X}, t) = \sum_{\alpha=1}^{M} \phi^{\alpha}(t)\zeta(\boldsymbol{X} - \boldsymbol{X}^{\alpha}) - U_R(\boldsymbol{X}) \tag{4.8}$$

where ϕ^{α} is atomistic potential, ϕ_X^{α} is the reference potential energy density of at 0 K, ζ is the localization function, and $U_R(\boldsymbol{X})$ is the reference potential energy density of atom α at 0 K.

Defining

$$\phi_I(\boldsymbol{X}_I, t) = \sum_{\alpha=1}^{M} \zeta_{I\alpha}(\boldsymbol{X}^{\alpha} - \boldsymbol{X}_I)\phi^{\alpha}(t) \tag{4.9}$$

$$\phi(\boldsymbol{X}, t) = \sum_{I=1}^{K} N_I(\boldsymbol{X})\phi_I(\boldsymbol{X}_I, t) \tag{4.10}$$

where $N_I(\boldsymbol{X})$ are the interpolation functions for the localization box, as defined in Section 4.5. Substituting Equations (4.9) and (4.10) in Equation (4.7), gives

$$U(\boldsymbol{X}, t) = \sum_{I=1}^{K}\sum_{\alpha=1}^{M} N_I(\boldsymbol{X})\zeta_{I\alpha}(\boldsymbol{X}^{\alpha} - \boldsymbol{X}_I)\phi^{\alpha}(t) - U_R(\boldsymbol{X}) \tag{4.11}$$

where M is the number of atoms in the localization box and $N_I(\boldsymbol{X})$ are interpolation functions, with $I = 1, K$, where K is the total number of nodes in the localization box

4.2.2 Evaluation of Displacement Gradient ($\underline{\boldsymbol{H}}$)

$$\underline{H}(\boldsymbol{X}, t) = \nabla_x \boldsymbol{u}(\boldsymbol{X}, t) = \nabla_x \sum_{I=1}^{K} N_I(\boldsymbol{X})\boldsymbol{u}_I(t) \tag{4.12}$$

$$\underline{H}(\boldsymbol{X},t)=\sum_{I=1}^{K}\sum_{\alpha=1}^{M}\nabla_{\boldsymbol{x}}N_I(\boldsymbol{X})\zeta_{I\alpha}(\boldsymbol{X}^{\alpha}-\boldsymbol{X}_I)\boldsymbol{u}^{\boldsymbol{\alpha}}(t) \tag{4.13}$$

where $\boldsymbol{u}(\boldsymbol{X},t)$ is the displacement vector of atom α, M is the number of atoms in the localization box and $N_I(\boldsymbol{X})$ are interpolation functions in the localization box, defined in Section 4.5.

4.2.3 Evaluation of First Piola-Kirchhoff Stress

The bond function is used to account for the interatomic forces, where one atom is inside the localization box, and the other one is outside. In such cases, only the fraction of the bond length inside the localization box is used for force computations. The bond function can be defined as,

$$B^{\alpha\beta}=\int_0^1\zeta\big(\lambda(\boldsymbol{X}_{\boldsymbol{\alpha}}-\boldsymbol{X})+(1-\lambda)(\boldsymbol{X}_{\boldsymbol{\beta}}-\boldsymbol{X})\big)d\lambda \tag{4.14}$$

where λ goes from 0 to 1 if β is inside the box. Otherwise, only till the fraction of $\boldsymbol{X}^{\alpha\beta}$ inside the box.

Setting $\boldsymbol{r}=\lambda\boldsymbol{X}^{\alpha\beta}+\boldsymbol{X}^{\beta}-\boldsymbol{X}\Rightarrow\boldsymbol{dr}=-\boldsymbol{dX}$ and taking the derivative of $\zeta\big(\lambda\boldsymbol{X}^{\alpha\beta}+\boldsymbol{X}^{\beta}-\boldsymbol{X}\big)$ with respect to λ,

$$\frac{\partial\zeta}{\partial\lambda}=\left(\frac{\partial\zeta}{\partial r}\right)\left(\frac{\partial r}{\partial\lambda}\right)=\left(-\frac{\partial\zeta}{\partial X}\right)(\boldsymbol{X}^{\alpha\beta})=-\boldsymbol{X}^{\alpha\beta}\cdot\nabla_{\boldsymbol{x}}\zeta\big(\lambda\boldsymbol{X}^{\alpha\beta}+\boldsymbol{X}^{\beta}-\boldsymbol{X}\big) \tag{4.15}$$

Integrating the above equation with respect to λ from 0 to 1,

$$\int_0^1\frac{\partial\zeta}{\partial\lambda}d\lambda=-\boldsymbol{X}^{\alpha\beta}\cdot\nabla_{\boldsymbol{x}}\int_0^1\zeta\big(\lambda\boldsymbol{X}^{\alpha\beta}+\boldsymbol{X}^{\beta}-\boldsymbol{X}\big)d\lambda \tag{4.16}$$

$$\zeta(\boldsymbol{X}^{\alpha}-\boldsymbol{X})-\zeta\big(\boldsymbol{X}^{\beta}-\boldsymbol{X}\big)=-\boldsymbol{X}^{\alpha\beta}\cdot\nabla_{\boldsymbol{x}}B^{\alpha\beta}(\boldsymbol{X}) \tag{4.17}$$

From the definition of momentum density,

$$\rho_0\frac{d\boldsymbol{V}}{dt}=\sum_{\alpha=1}^{M}m^{\alpha}\frac{d\boldsymbol{V}^{\alpha}}{dt}\zeta(\boldsymbol{X}^{\alpha}-\boldsymbol{X})=\sum_{\alpha=1}^{M}(\boldsymbol{f}^{\alpha}+m^{\alpha}\boldsymbol{b}^{\alpha})\psi(\boldsymbol{X}^{\alpha}-\boldsymbol{X}) \tag{4.18}$$

From internal force term,

$$\begin{aligned}\sum_{\alpha=1}^{M}\boldsymbol{f}^{\alpha}\zeta(\boldsymbol{X}^{\alpha}-\boldsymbol{X})&=\sum_{\alpha=1}^{M}\sum_{\beta\neq\alpha}^{M}\boldsymbol{f}^{\alpha\beta}\zeta(\boldsymbol{X}^{\alpha}-\boldsymbol{X})\\&=\frac{1}{2}\sum_{\alpha=1}^{M}\sum_{\beta\neq\alpha}^{M}\boldsymbol{f}^{\alpha\beta}\big[\zeta(\boldsymbol{X}^{\alpha}-\boldsymbol{X})-\zeta\big(\boldsymbol{X}^{\beta}-\boldsymbol{X}\big)\big]\end{aligned} \tag{4.19}$$

Substituting Equations (4.17) and (4.19) in Equation (4.18),

$$\rho_0 \frac{d\boldsymbol{V}}{dt} = \sum_{\alpha=1}^{M} \left\{ \frac{1}{2} \sum_{\beta \neq \alpha}^{M} \boldsymbol{f}^{\alpha\beta} \left[-\boldsymbol{X}^{\alpha\beta} \cdot \nabla_{\boldsymbol{x}} B^{\alpha\beta}(\boldsymbol{X}) \right] + m^{\alpha} \boldsymbol{b}^{\alpha} \zeta(\boldsymbol{X}^{\alpha} - \boldsymbol{X}) \right\} \tag{4.20}$$

$$\rho_0 \frac{d\boldsymbol{V}}{dt} = \nabla_{\boldsymbol{x}} \cdot \left\{ -\frac{1}{2} \sum_{\alpha=1}^{M} \sum_{\beta \neq \alpha}^{M} \boldsymbol{f}^{\alpha\beta} \bigotimes \boldsymbol{X}^{\alpha\beta} \boldsymbol{B}^{\alpha\beta}(X) \right\} + \sum_{\alpha=1}^{M} m^{\alpha} \boldsymbol{b}^{\alpha} \zeta(\boldsymbol{X}^{\alpha} - \boldsymbol{X}) \tag{4.21}$$

From momentum balance equation,

$$\rho_0 \frac{d\boldsymbol{V}}{dt} = \nabla_{\boldsymbol{x}} \cdot \underline{P} + \rho_0 \boldsymbol{b} \tag{4.22}$$

Comparing Equations (4.21) and (4.22),

$$\underline{P}(\boldsymbol{X}, t) = -\frac{1}{2} \sum_{\alpha=1}^{M} \sum_{\beta \neq \alpha}^{M} \boldsymbol{f}^{\alpha\beta} \bigotimes \boldsymbol{X}^{\alpha\beta} B^{\alpha\beta}(\boldsymbol{X}) \tag{4.23}$$

The expression for bond function is defined as follows. Considering a localization box as shown in Figure 4.2 with one atom (α) inside the box and another atom (β) outside the box. In order to correctly scale the influence of the atom outside the box, a new position of atom β is determined as,

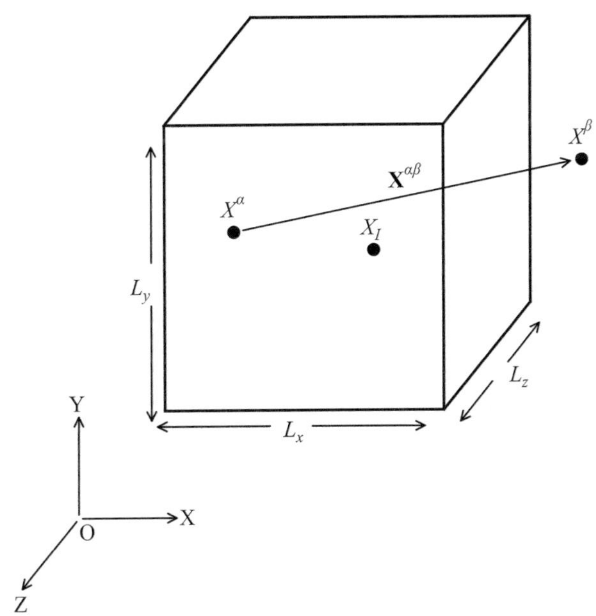

FIGURE 4.2 Illustration of a localization box with one atom inside and another outside the box.

$$\text{If } X_1^\beta \geq X_1^I + \frac{L_X}{2} \quad \text{then} \quad X_1^{\beta N} = X_1^I + \frac{L_X}{2} \quad \text{else} \quad X_1^{\beta N} = X_1^\beta$$

$$\text{If } X_1^\beta \leq X_1^I - \frac{L_X}{2} \quad \text{then} \quad X_1^{\beta N} = X_1^I - \frac{L_X}{2} \quad \text{else} \quad X_1^{\beta N} = X_1^\beta$$

$$\text{If } X_2^\beta \geq X_2^I + \frac{L_Y}{2} \quad \text{then} \quad X_2^{\beta N} = X_2^I + \frac{L_Y}{2} \quad \text{else} \quad X_2^{\beta N} = X_2^\beta$$

$$\text{If } X_2^\beta \leq X_2^I - \frac{L_Y}{2} \quad \text{then} \quad X_2^{\beta N} = X_2^I - \frac{L_Y}{2} \quad \text{else} \quad X_2^{\beta N} = X_2^\beta$$

$$\text{If } X_3^\beta \geq X_3^I + \frac{L_Z}{2} \quad \text{then} \quad X_3^{\beta N} = X_3^I + \frac{L_Z}{2} \quad \text{else} \quad X_3^{\beta N} = X_3^\beta$$

$$\text{If } X_3^\beta \leq X_3^I - \frac{L_Z}{2} \quad \text{then} \quad X_3^{\beta N} = X_3^I - \frac{L_Z}{2} \quad \text{else} \quad X_3^{\beta N} = X_3^\beta$$

where the superscript "N" denotes the new value of the position and L_x, L_y, L_z are the dimensions of the localization box, and I is the centroid of the localization box. Now,

$$\text{If } |X_i^\beta - X_i^\alpha| \leq 10^{-3} \quad \text{then} \quad R_i = 1 \quad \text{else}$$

$$R_1 = \frac{\left(X_1^{\beta N} - X_1^\alpha\right)}{\left(X_1^\beta - X_1^\alpha\right)}$$

$$R_2 = \frac{\left(X_2^{\beta N} - X_2^\alpha\right)}{\left(X_2^\beta - X_2^\alpha\right)}$$

$$R_3 = \frac{\left(X_3^{\beta N} - X_3^\alpha\right)}{\left(X_3^\beta - X_3^\alpha\right)}$$

$$\text{If } R_1 \leq R_2 \quad \text{then} \quad R = R_1, \quad \text{else} \quad R = R_2. \text{ If } R_3 \leq R \quad \text{then} \quad R = R_3$$

Then, bond function is given by

$$B^{\alpha\beta} = \frac{R}{L_X L_Y L_z} \tag{4.24}$$

4.3 ATOMISTIC J-INTEGRAL AT FINITE TEMPERATURE: ENTROPIC EFFECT

From statistical mechanics, the Helmholtz free energy density is given by [7],

$$\Psi = U - T\eta = -\frac{k_B T}{V}\text{Log}\, Z \tag{4.25}$$

where U is the internal energy density, T is the temperature, η is the entropy density, V is the volume of the ensemble, k_B is Boltzmann's constant, and Z is the partition function of the atoms occupying the region. Note that the general definition of free energy density Ψ includes the entropy term and therefore is valid for finite temperature applications of the atomistic J-integral. The conventional definition of the J-integral under isothermal conditions does not take into account the entropic contribution to the free energy, and consequently, may lead to a significant overestimation of the J-integral at the atomistic level at elevated temperatures. As alluded to in Equation (4.1), the total J-integral at a finite temperature is given by

$$J_T = J_U - J_\eta \tag{4.26}$$

In Equation (4.26), J_U is the value of the J-integral without considering the entropic contribution and is given by Equation (4.2), and J_η is the entropic contribution. It is evident that at 0 K, the $T\eta$ term in Equation (4.25) is absent and hence, the equation for J_U can be used to compute J_T. However, at higher temperatures, the $T\eta$ term can have a significant contribution due to thermal excitation of the atoms leading to a reduction in J_T. Based on the Cauchy-Born rule, entropic contribution J_η can be quantified for a defect-free crystalline material using the local harmonic (LH) approximation and is given by Equation (4.27) ([4]).

$$J_\eta = \int_{\partial\Omega} \left\langle \frac{k_B T}{V_\alpha} \log\left(\left(\frac{\hbar}{k_B T}\right)^3 \sqrt{\det D_{\text{LH}}} \right) \right\rangle N dA \tag{4.27}$$

In Equation (4.27), k_B is Boltzmann's constant, $\hbar$ is Planck's constant, V_α is volume of the atom, T is the absolute temperature, and D_{LH} is the dynamical matrix based on the LH approximation of atoms vibrating within a defect-free crystal lattice. The LH approximation is essentially an Einstein model of the vibrational frequencies and has been used extensively in MD-continuum coupling. As shown in Equation (4.27), computation of the dynamical matrix is necessary to quantify the entropic contribution to the J-integral. At elevated temperatures, there is significant random thermal vibration of the atoms about their mean position (not to be confused with the translational motion of atoms due to propagating stress waves) and the vibrational modes of the lattice are given by the eigenvalues of the dynamical matrix. The procedure to compute dynamical matrix is discussed in the next section.

4.3.1 Evaluation of Dynamical Matrix (D_{LH})

From statistical mechanics, the dynamical matrix required for quantifying the entropic contribution to the J-integral is given by Jones et al. [4],

$$D_{\mathrm{LH}} = \frac{1}{m}\frac{\partial^2 \Phi}{\partial u_0^2} = \frac{1}{m}\frac{\partial}{\partial u_0}\sum_{\beta \neq 0} \phi'(r_{\beta 0})\frac{\overline{r_{\beta 0}}}{r_{\beta 0}} \tag{4.28}$$

where $\Phi = \frac{1}{2V}\sum_{\alpha}^{N}\sum_{\beta=0}^{N}\phi(r_{\alpha\beta})$, where $r_{\alpha\beta}$ is the interatomic distance between atoms α and β, u_0 is the displacement vector, m is the atomic mass, $\phi(r)$ is the pairwise potential, V is the volume, $r_{\beta 0}$ is the position vector of the atom measured from the origin of the MD box. Using the chain-rule of differentiation, Equation (4.28) can be expressed as,

$$D_{\mathrm{LH}} = \frac{1}{m}\sum_{\beta \neq 0}\left(\phi''(r_{\beta 0})\frac{\overline{r_{\beta 0}}}{r_{\beta 0}} \otimes \frac{\overline{r_{\beta 0}}}{r_{\beta 0}} + \phi'(r_{\beta 0})\frac{1}{r_{\beta 0}}\left[I - \frac{\overline{r_{\beta 0}}}{r_{\beta 0}} \otimes \frac{\overline{r_{\beta 0}}}{r_{\beta 0}}\right]\right) \tag{4.29}$$

where I is the identity tensor. Using data from MD simulations in Equation (4.29), the dynamical matrix can be evaluated, which can then be used to compute the entropic contribution to the J-integral using Equation (4.27). For example, consider a pairwise potential given by,

$$\phi(r) = k(r - r_0)^2 \tag{4.30}$$

where k is the energy constant and r_0 is the equilibrium interatomic distance.

The position vector and its magnitude for an atom β are given by,

$$\overline{r_\beta} = x_\beta \hat{i} + y_\beta \hat{j} + z_\beta \hat{k} \tag{4.31}$$

$$r_\beta^2 = x_\beta^2 + y_\beta^2 + z_\beta^2 \tag{4.32}$$

The terms required for evaluation of D_{LH} can then be written as,

$$\frac{\overline{r_\beta}}{r_\beta} \otimes \frac{\overline{r_\beta}}{r_\beta} = \begin{bmatrix} \frac{x_\beta^2}{r_\beta^2} & \frac{x_\beta y_\beta}{r_\beta^2} & \frac{x_\beta z_\beta}{r_\beta^2} \\ \frac{x_\beta y_\beta}{r_\beta^2} & \frac{y_\beta^2}{r_\beta^2} & \frac{y_\beta z_\beta}{r_\beta^2} \\ \frac{x_\beta z_\beta}{r_\beta^2} & \frac{y_\beta z_\beta}{r_\beta^2} & \frac{z_\beta^2}{r_\beta^2} \end{bmatrix} \tag{4.33}$$

$$I-\frac{\overline{r_\beta}}{r_\beta}\bigotimes\frac{\overline{r_\beta}}{r_\beta}=\begin{bmatrix}1-\frac{x_\beta^2}{r_\beta^2} & -\frac{x_\beta y_\beta}{r_\beta^2} & -\frac{x_\beta z_\beta}{r_\beta^2}\\ -\frac{x_\beta y_\beta}{r_\beta^2} & 1-\frac{y_\beta^2}{r_\beta^2} & -\frac{y_\beta z_\beta}{r_\beta^2}\\ -\frac{x_\beta z_\beta}{r_\beta^2} & -\frac{y_\beta z_\beta}{r_\beta^2} & 1-\frac{z_\beta^2}{r_\beta^2}\end{bmatrix} \tag{4.34}$$

$$\phi'(r)\frac{1}{r_\beta}\left[I-\frac{\overline{r_\beta}}{r_\beta}\bigotimes\frac{\overline{r_\beta}}{r_\beta}\right]=\frac{2k\left(r_\beta-r_0\right)}{r_\beta}\begin{bmatrix}1-\frac{x_\beta^2}{r_\beta^2} & -\frac{x_\beta y_\beta}{r_\beta^2} & -\frac{x_\beta z_\beta}{r_\beta^2}\\ -\frac{x_\beta y_\beta}{r_\beta^2} & 1-\frac{y_\beta^2}{r_\beta^2} & -\frac{y_\beta z_\beta}{r_\beta^2}\\ -\frac{x_\beta z_\beta}{r_\beta^2} & -\frac{y_\beta z_\beta}{r_\beta^2} & 1-\frac{z_\beta^2}{r_\beta^2}\end{bmatrix} \tag{4.35}$$

$$\phi''(r)\frac{\overline{r_\beta}}{r_\beta}\bigotimes\frac{\overline{r_\beta}}{r_\beta}=\begin{bmatrix}\frac{2kx_\beta^2}{r_\beta^2} & \frac{2kx_\beta y_\beta}{r_\beta^2} & \frac{2kx_\beta z_\beta}{r_\beta^2}\\ \frac{2kx_\beta y_\beta}{r_\beta^2} & \frac{2ky_\beta^2}{r_\beta^2} & \frac{2ky_\beta z_\beta}{r_\beta^2}\\ \frac{2kx_\beta z_\beta}{r_\beta^2} & \frac{2ky_\beta z_\beta}{r_\beta^2} & \frac{2kz_\beta^2}{r_\beta^2}\end{bmatrix} \tag{4.36}$$

Substituting Equations (4.35) and (4.36) in Equation (4.29), the final for evaluation of D_{LH} can then be written as,

$$D_{\text{LH}}=\frac{1}{m}\sum_{\beta\neq 0}\left(\begin{bmatrix}\frac{2kx_\beta^2}{r_\beta^2} & \frac{2kx_\beta y_\beta}{r_\beta^2} & \frac{2kx_\beta z_\beta}{r_\beta^2}\\ \frac{2kx_\beta y_\beta}{r_\beta^2} & \frac{2ky_\beta^2}{r_\beta^2} & \frac{2ky_\beta z_\beta}{r_\beta^2}\\ \frac{2kx_\beta z_\beta}{r_\beta^2} & \frac{2ky_\beta z_\beta}{r_\beta^2} & \frac{2kz_\beta^2}{r_\beta^2}\end{bmatrix}+\frac{2k\left(r_\beta-r_0\right)}{r_\beta}\begin{bmatrix}1-\frac{x_\beta^2}{r_\beta^2} & -\frac{x_\beta y_\beta}{r_\beta^2} & -\frac{x_\beta z_\beta}{r_\beta^2}\\ -\frac{x_\beta y_\beta}{r_\beta^2} & 1-\frac{y_\beta^2}{r_\beta^2} & -\frac{y_\beta z_\beta}{r_\beta^2}\\ -\frac{x_\beta z_\beta}{r_\beta^2} & -\frac{y_\beta z_\beta}{r_\beta^2} & 1-\frac{z_\beta^2}{r_\beta^2}\end{bmatrix}\right) \tag{4.37}$$

Defining $k = 469$ Kcal/$\left(\text{mole Å}^2\right)$, $r_0 = 1.40$ Å, valid for Optimized Potentials for Liquid Simulations (OPLS) potential for a single graphene sheet at 300 K, a sample numerical value for computation of D_{LH} is shown below in Equation (4.38).

$$D_{\text{LH}} = \begin{bmatrix} 4.5956 & 4.3837 & 0 \\ 4.3837 & 4.2034 & 0 \\ 0 & 0 & 0.0114 \end{bmatrix} \times 10^7 \text{ Kcal}/\left(\text{mole } \text{Å}^2\right) \tag{4.38}$$

4.4 COHESIVE CONTOUR BASED APPROACH FOR *J*-INTEGRAL

As mentioned earlier, the entropic contribution (due to $T\eta$ term in Equation (4.25)) can have a significant effect on the computed J-integral value, because that portion of the free energy goes into thermal expansion and, therefore, is not available for crack growth. As defined in Equation (4.26), the total atomistic J-integral at finite temperature (J_T) is given by $J_T = J_U - J_\eta$, where J_U is the contribution to the J-integral due to internal energy (U) computed from Equation (4.2), and J_η is the entropic contribution (Equation (4.27)). As discussed in Jones et al. [4], this methodology is only applicable to a defect-free crystalline structure and hence cannot be readily applied to amorphous materials, such as polymers.

However, the path independence of the J-integral facilitates the use of a cohesive-contour based technique to compute the J-integral, as described in Klein et al. [8]. A J-integral contour in the shape of a narrow strip encompassing the thin cohesive contour ahead of the crack tip is constructed, as illustrated in Figure 4.3. Because of the way the contour is designed, it can be shown that, by making the contour width (h) sufficiently small, the Helmholtz free energy term can be ignored from the J-integral equation for Mode-I as well as for mixed-mode loadings. If we consider the scalar (indicial) form of the expression for the total J-integral,

$$(J_T)_k = \int_\Gamma \langle \Psi \delta_{kj} - P_{ij} H_{ik} \rangle N_j d\Gamma \tag{4.39}$$

where Ψ is the Helmholtz free energy density, δ_{kj} is the Kronecker delta, and repeated indices imply summation. If we set $k = 1$ to obtain $(J_T)_1$, and $j = 2$ to define the normal N_2 along the narrow cohesive-contour in Figure 4.3, the first term on the RHS of Equation (4.39) vanishes because $\delta_{12} = 0$ by definition. Therefore, the only terms that need to be evaluated along the narrow cohesive contour are the

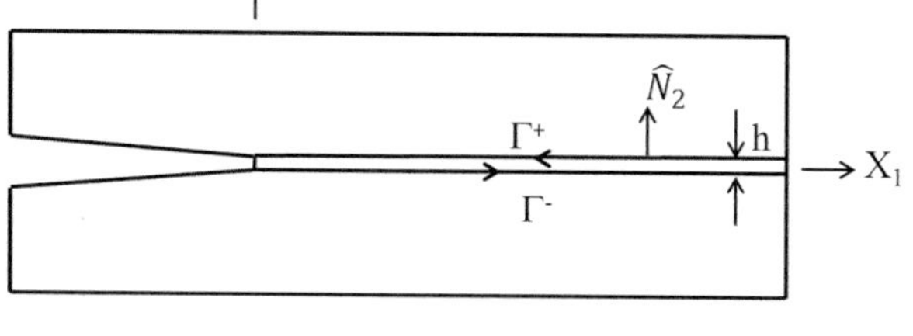

FIGURE 4.3 J-integral contour in the form of a narrow strip ahead of crack tip.

Cartesian components of the first Piola-Kirchhoff stress tensor (P_{i2}), and components of the displacement gradient tensor (H_{i1}), using Equations (4.6) and (4.5), respectively, giving

$$(J_T)_1 = \int_{\Gamma} \langle -P_{i2} H_{i1} \rangle N_2 d\Gamma \tag{4.40}$$

This approach is especially useful for thermoset polymers since the Cauchy-Born rule and the LH approximation-based approach in Equation (4.27) cannot be used for quantifying the entropic contribution for amorphous (disordered) thermoset polymers. It should be noted that the expression in Equation (4.40) is an approximation to the analytical value for J_T, and convergence studies will need to be carried out to determine how narrow the height of the cohesive contour, h, needs to be to provide a sufficiently accurate value of J for an amorphous material, such as polymers. A detailed case study of this cohesive contour approach is presented in Section 7.2. It is also essential that the fully developed fracture process zone is contained within the narrow cohesive contour at crack initiation, which might be challenging for a polymer with a large process zone size.

4.5 NUMERICAL EVALUATION OF ATOMISTIC J-INTEGRAL

Numerical integration through Gaussian quadrature was employed to evaluate the atomistic J-integral using the equations given in previous sections. The J-integral contour around the crack is divided into segments, and each segment is further divided into localization boxes, as shown in Figure 4.4(a) and (b). The discrete atomistic values of potentials and displacement gradients obtained from MD simulations were converted into field quantities using the finite-element-type interpolation functions for a nine-node element as given by Equation (4.41), in terms of local co-ordinates ξ and χ (i.e., set $K = 9$ in Equations (4.4) and (4.5)). The corresponding node numbering is shown in Figure 4.4(b).

$$\begin{aligned}
N_1 &= \frac{1}{4}(1-\xi)(1-\chi)(\xi\chi) \\
N_2 &= -\frac{1}{4}(1+\xi)(1-\chi)(\xi\chi) \\
N_3 &= \frac{1}{4}(1+\xi)(1+\chi)(\xi\chi) \\
N_4 &= -\frac{1}{4}(1-\xi)(1+\chi)(\xi\chi) \\
N_5 &= -\frac{1}{2}(1-\xi^2)(1-\chi)(\chi) \\
N_6 &= \frac{1}{2}(1+\xi)(1-\chi^2)(\xi) \\
N_7 &= \frac{1}{2}(1-\xi^2)(1+\chi)(\chi) \\
N_8 &= -\frac{1}{2}(1-\xi)(1-\chi^2)(\xi) \\
N_9 &= (1-\xi^2)(1-\chi^2)
\end{aligned} \tag{4.41}$$

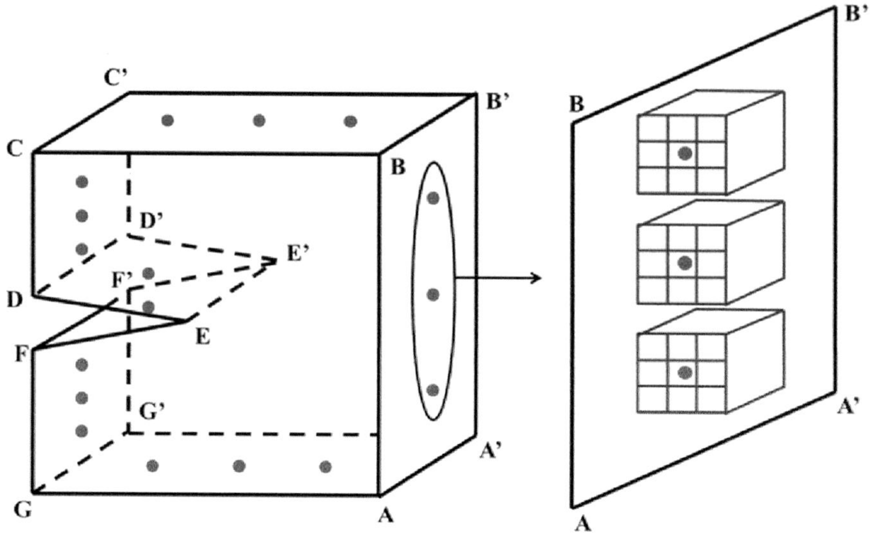

FIGURE 4.4 Gaussian quadrature points along the J-integration contour depicting localization boxes on one side. (a) Side view of the contour. (b) Top view of the contour.

J-integral for the side AB is given by,

$$J_{AB} = h\int_a^b \left(U\hat{N} - \underline{H}^T\,\underline{P}\hat{N}\right)dY \tag{4.42}$$

where the integration limits $Y = a$ and $Y = b$ are shown in Figure 4.4(b), and h is the depth of the contour, given by AA' in Figure 4.4.

4.5.1 Gaussian Quadrature

The integration of the J-integral, due to its mathematical complexity, has to be carried out using numerical integration, in this case, Gaussian quadrature. Because Gaussian quadrature requires the limits of the integral to be from -1 to 1, a change of variables is required. The following steps illustrate how the integration is carried out along the contour path AB in Figure 4.4(a).

Referring to Figure 4.4(b), we assume a local to global linear transformation $Y = C_1\xi + C_2$, where the local coordinate ξ ranges from -1 to 1, and C_1 and C_2 are constants.

Applying boundary conditions from Figure 4.4(b) that at $\xi = -1, Y = a$, hence $a = C_1(-1) + C_2$

Similarly, at $\xi = 1, Y = b$, and hence $b = C_1(1) + C_2$, which leads to $C_1 = \frac{b-a}{2}$ and $C_2 = \frac{b+a}{2}$ and

TABLE 4.1
Sample Values of *J*-Integral Computed Along a Closed Contour Path *ABCDEFGA* for Graphene at *T*=0 K with OPLS Potential

	J_I (J/m^2)					
Applied *K* (MPa m$^{1/2}$)	***AB***	***CD***	***FG***	***BC***	***GA***	**Total**
0	0	0	0	0	0	0
1.22	1.2369	0.6075	0.778	−0.4306	−0.4305	1.7613
2.44	5.0363	2.526	2.7425	−1.8107	−2.1275	6.3666
3.67	11.0652	5.7634	6.087	−4.2981	−4.3981	14.2194

$$dY = C_1 d\xi = \left(\frac{b-a}{2}\right) d\xi$$

Equation (4.42) can now be converted to a form suitable for numerical integration in terms of local coordinate ξ as shown below,

$$\begin{aligned} \boldsymbol{J}_{AB} &= h \int_{-1}^{1} \left(U\hat{N} - \underline{H}^{\mathrm{T}}\,\underline{P}\hat{N}\right) C_1 d\xi \\ \text{Or}, \boldsymbol{J}_{AB} &= C_1 h \sum_{k=1}^{N_G} \left[U(\xi_k)\hat{N} - \underline{H}^{\mathrm{T}}(\xi_k)\underline{P}(\xi_k)\hat{N}\right] w_k \\ \text{Or}, \boldsymbol{J}_{AB} &= C_1 h \sum_{k=1}^{N_G} \left[U(Y_k)\hat{N} - \underline{H}^{\mathrm{T}}(Y_k)\underline{P}(Y_k)\hat{N}\right] w_k \end{aligned} \tag{4.43}$$

where, $Y_k = C_1\xi_k + C_2$, and N_G are the number of Gauss points, and w_k are the Gaussian weights, and h is the depth of the contour.

Therefore, the J-integral can be calculated using this procedure at all Gaussian quadrature points and for all segments of the contour. Referring to Figure 4.4, sample values of the J-integral computed for a graphene sheet at 0 K for each section of the path are shown in Table 4.1. Details regarding the graphene sheet used, simulation parameters, and more results are presented in Section 4.6.

4.6 CASE STUDY 1: ATOMISTIC *J*-INTEGRAL CALCULATION FOR A CENTER-CRACKED GRAPHENE SHEET USING OPLS

A single monolayer graphene sheet with a center crack was modeled using MD (LAMMPS software) and was subjected to a far field uniaxial stress in the Y-direction (pure Mode-I loading). The atomistic J-integral was computed on concentric contours as shown in Figure 4.5.

In this example, the graphene sheet was modeled using harmonic style for bonds and angles, and OPLS style for dihedrals. The graphene sheet, measuring

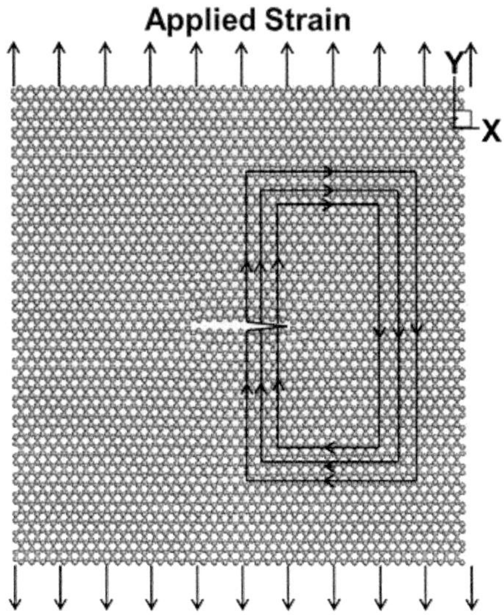

FIGURE 4.5 Graphene sheet with nanoscale crack and contours for J-integral computation.

10.5 nm × 10 nm × 0.34 nm, was made up of 4420 carbon atoms in armchair configuration. The initial length of the pre-existing (starter) crack was 2.2 nm in the graphene sheet as shown in Figure 4.5, which is greater than the critical crack length for graphene. The coarse time step for this simulation (Δt) was set equal to 0.1 fs, and the applied strain rate was 2×10^{11}/s.

In order to establish proof of concept for the atomistic J-integral computation, the initial MD simulations were carried out at temperature $T = 0$ K, i.e., when there is no entropic contribution to the free energy due to thermal motion in accordance with Equation (4.25). Pressure barostatting was not used in these simulations. A convergence study was conducted on the effect of localization box size on the computed values of virial stress in the graphene sheet away from the crack tip. It was found that the computed value of stress becomes box-size independent for box sizes greater than 2 nm (~20 Å) for monolayer graphene, as shown in the stress convergence plot in Figure 4.6.

The atomistic J-integral results are shown as a function of applied Mode-I stress intensity factor (K_I) in Figure 4.7 for a graphene sheet and compared with LEFM predictions.

The plot shows computed values for the three paths around the crack for a localization box-size of 27 Å. Figure 4.7 shows that the quadratic dependence of the atomistic J-integral on Mode-I stress intensity factor (K_I) is in good agreement with plane-stress LEFM predictions for pure Mode-I loading, i.e., $J_I = K_I^2/E$, where E is the Young's modulus $(E = 0.85$ Tpa) of graphene at 0 K obtained from our MD simulations. The stress intensity factor can be computed from the virial stress values obtained from MD output. The equation for computation of stress intensity factor

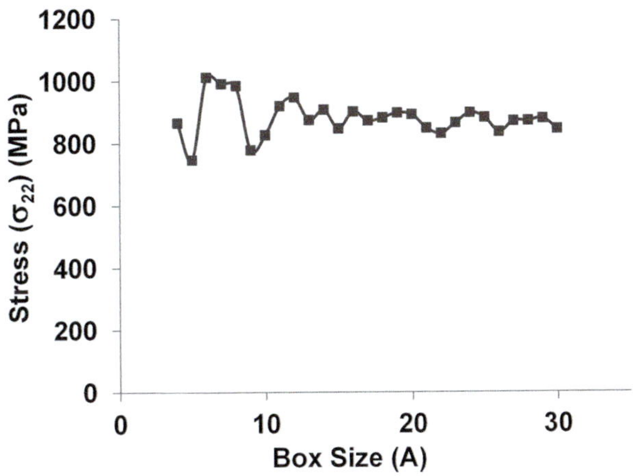

FIGURE 4.6 Influence of localization box size on computed stress values in graphene.

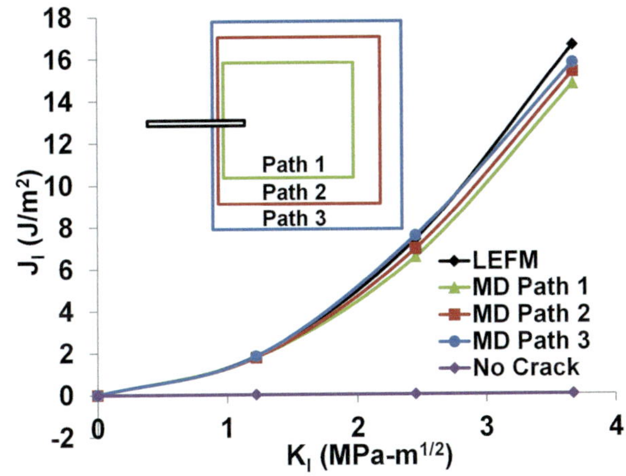

FIGURE 4.7 Atomistic J_I vs. K_I for a graphene platelet at 0 K with harmonic potential with 27 Å box size compared with LEFM.

(K) can be written in terms of virial stress (σ), half crack length (a), and geometry factor (f)) as $K_I = \sigma f \sqrt{\pi a}$.

Figure 4.7 also verifies that J is identically zero in the absence of a singularity within the integration contour, as depicted by the "no crack" line in the figure. More importantly, it is clear from Figure 4.7 that the calculated value of J_I is reasonably path-independent for the graphene platelet, even at the nanoscale. This result facilitates the use of calculating the J-integral at finite temperatures using a cohesive-contour technique discussed earlier.

4.7 CASE STUDY 2: ATOMISTIC *J*-INTEGRAL CALCULATION FOR A CENTER-CRACKED NANOGRAPHENE PLATELET AT FINITE TEMPERATURE USING OPLS (T=300 K)

Atomistic J-integral was computed for a center-cracked graphene platelet at temperatures of 0.1, 100, 200, and 300 K to verify the model at finite temperatures and quantify the entropic contribution to J. The center-cracked graphene platelet used is the same as the one used for 0 K computations in Section 4.6 and as shown in Figure 4.5. The system temperature was elevated to 300 K and was allowed to equilibrate at that temperature and at a pressure of zero atmospheres using the NPT fix command in LAMMPS. As in Section 4.6, the graphene sheet was modeled using harmonic style for bonds and angles, and OPLS style for dihedrals. Poisson's effect in graphene was considered in this simulation, and a value of 0.22 was used for the Poisson's ratio, which is in good agreement with literature, as given by Zhao and Shi [9]. The graphene platelet was stretched under Mode-I tension, as was the case for 0 K, and the resulting data from the MD simulation was used to compute the atomistic J-integral. Three different box sizes, 22, 27, and 32 Å were used to compute the J-integral at temperature $T = 0.1, 100, 200,$ and 300 K (averaged over five concentric paths) using the techniques discussed earlier, and the results were found to be independent of the box size. The results of one of these computations at $T = 0.1$ K are shown in Figure 4.8, with the symbols representing the mean value and the error bar representing the standard deviation based on averaging over five concentric contours.

J_U is the portion of the atomistic J-integral which includes only internal energy contribution, (i.e., without entropic contribution) averaged over five concentric paths. As one might expect, it shows good agreement with strain-energy-based LEFM results except at low temperatures. J_T is the total computed J-integral including both energetic and entropic contributions (again, averaged over five paths). Figure 4.8 also shows J_{cohesive} computed using the novel cohesive-contour technique discussed earlier, and it is in good agreement with J_T.

It is interesting to note that at $T = 0.1$ K, the energetic component of J (J_U) is identical to the total J, (J_T), due to the absence of entropic effects at such low temperature, as is evident in Figure 4.8. Since most experiments are conducted at room temperature, results for $T = 300$ K case will be discussed in detail. Figure 4.9 shows the atomistic J-integral compared with the LEFM prediction using a localization box size of 27 Å.

As before, J_U is the portion of the atomistic J-integral which includes only internal energy contribution, (i.e., without entropic contribution) averaged over five concentric paths. As one might expect, it shows good agreement with strain-energy-based LEFM results. J_T is the total computed J-integral including both energetic and entropic contributions (again, averaged over five paths), and it is evident that it is significantly (~50%) lower than LEFM predictions, in accordance with Equation (4.25). Figure 4.9 also shows J_{cohesive} computed using the novel cohesive-contour technique discussed in Chapter 4, and it is in good agreement with J_T. This is an important benchmark result, as it potentially enables the use of cohesive-contour based technique for computing J-integral at the nanoscale for the case of amorphous

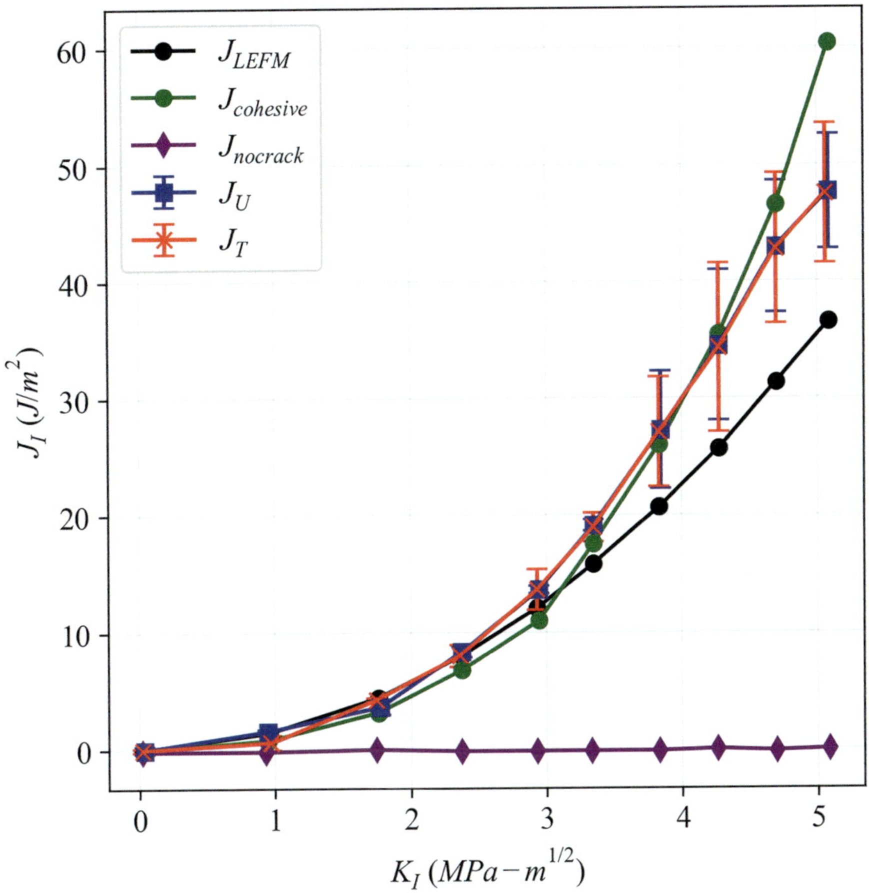

FIGURE 4.8 Atomistic J_I vs. K_I for a graphene platelet at $T = 0.1$ K with harmonic potential and 27 Å localization box size using OPLS.

materials, where direct computation of entropic contribution using the LH approximation approach is no longer feasible. However, it must be verified that the fully developed fracture process zone for the amorphous material is included in the cohesive contour for accurate estimation of toughness, as it was in the case of monolayer graphene.

4.8 CASE STUDY 3: ATOMISTIC J-INTEGRAL CALCULATION FOR A CENTER-CRACKED NANOGRAPHENE PLATELET USING REAXFF

OPLS based on the harmonic force field were employed in all the molecular dynamics (MD) simulations presented in the previous sections. While the OPLS force field works well for evaluating mechanical properties, it cannot simulate

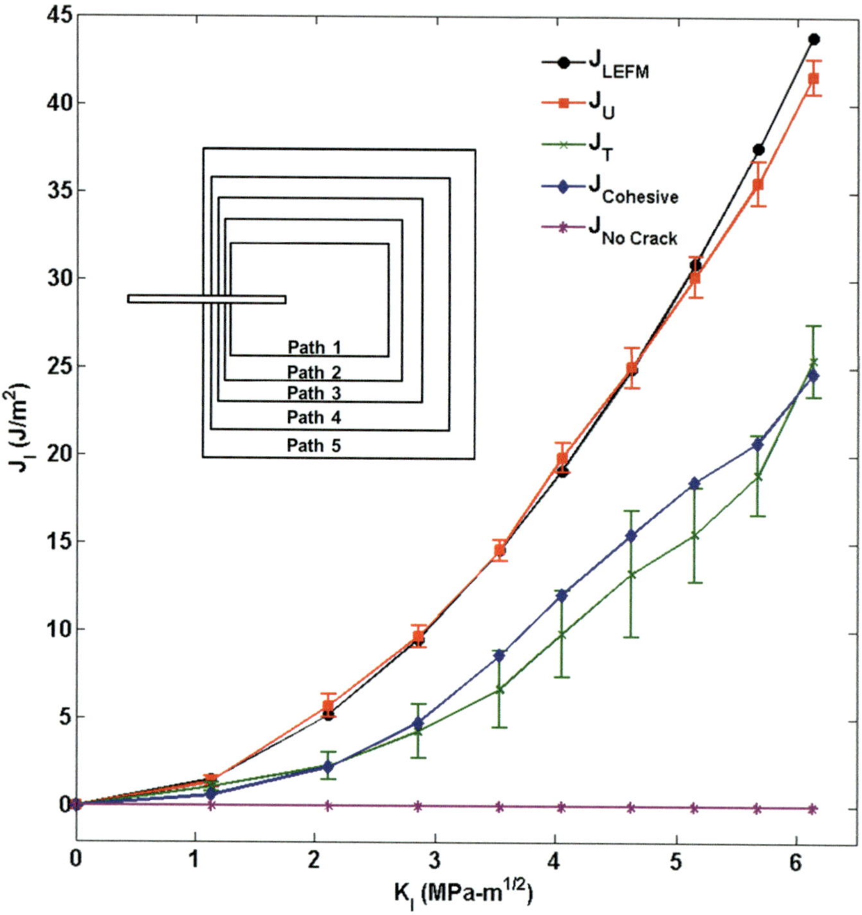

FIGURE 4.9 Atomistic J_I vs.K_I for a graphene platelet at $T = 300$ K with harmonic potential with 27 Å localization box size using OPLS.

phenomena such as bond formation and breakage. In order to better understand the fracture behavior of the material, it is important to be able to simulate crack initiation and propagation. For that purpose, a new force field, ReaxFF, was employed. ReaxFF is a bond order-based force field which allows for continuous bond formation/breaking, as described in Van Duin et al. [10]. Bonded interactions are generated on the fly, based on distance-dependent bond order functions.

As a benchmark case study, MD simulations were carried out on a center-cracked nano-graphene sheet using the ReaxFF force field. The potential parameters used in this simulation were obtained from [11]. The graphene sheet consists of 4398 atoms and is similar in dimensions to the one used in OPLS simulations i.e., 105 × 100 Å, in an armchair configuration. An initial central crack of approximately 20 Å was generated by deleting the atoms at the center of the graphene sheet. All the finite temperature simulations were done using the NPT fix command in order to

FIGURE 4.10 Center-cracked graphene at the beginning and end of fracture simulation using ReaxFF at $T = 300$ K.

fully account for Poisson's effect. A time step of 0.1 femtosecond and a strain rate of 3×10^{11}/s were used for the simulations. A total tensile strain of approximately 25% was gradually applied on the MD simulation box in the direction perpendicular to the crack to simulate Mode-I loading. The ReaxFF force field was able to successfully simulate the initiation and propagation of the center crack in the graphene sheet, as depicted in Figure 4.10.

For preliminary benchmarking, the atomistic J-integral was evaluated at 0.1 and 300 K using ReaxFF. Based on the results obtained for harmonic OPLS simulations for nanographene, it was established that for a computation box size above 20 Å, the J-integral computations are fairly independent of localization box size. Hence, a box size of 27 Å was selected for consistency between OPLS and ReaxFF computations.

Figure 4.11 shows the comparison of atomistic J-integral values computed at 0.1 K using various methods (discussed in previous sections) compared to the values from LEFM for a localization box size of 27 Å (for one contour). Figure 4.12 shows the comparison of atomistic J-integral values computed at 300 K using different methods compared to the values from LEFM for a localization box size of 27 Å (averaged over three concentric path-independent contours).

In these plots, J_U is the portion of the atomistic J-integral which includes only internal energy contribution, (i.e., without entropic contribution), and it shows good agreement with purely strain-energy-based LEFM results. J_T is the total computed J-integral including both energetic and entropic contributions, calculated using the partition function approach, as discussed in Jones et al. [4]. J_{cohesive} is the J-integral value computed using the narrow cohesive contour approach discussed earlier. There is good agreement between LEFM, the energetic J-integral (J_U), and the total J-integral (J_T) at 0.1 K as depicted in Figure 4.11, because at 0.1 K the entropic

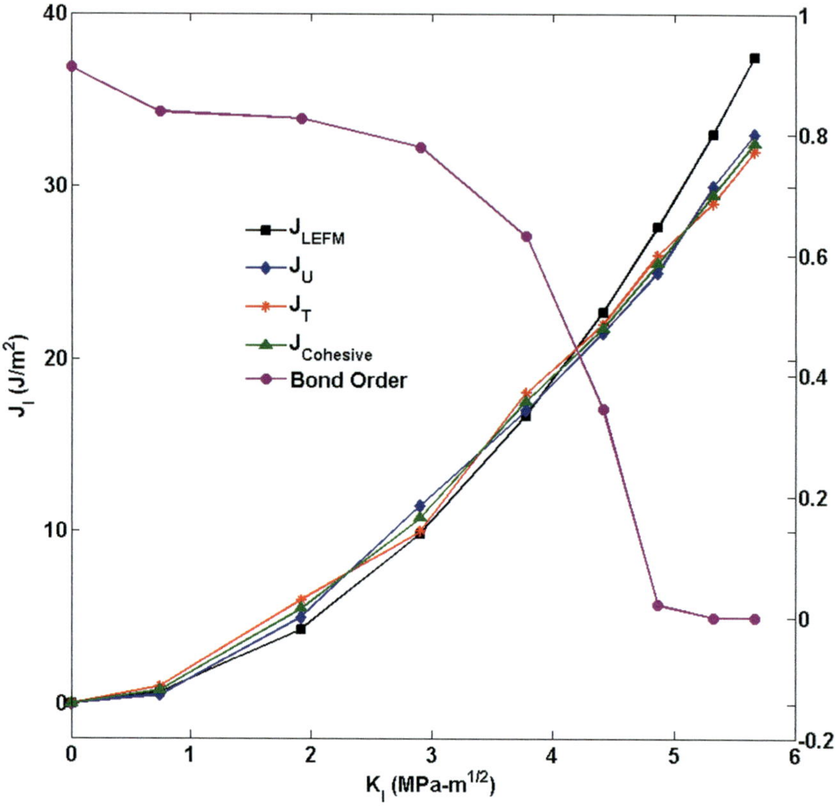

FIGURE 4.11 J-integral computation for graphene as a function of K_{I} using ReaxFF at $T = 0.1$ K; also shown is change in bond order with loading (27 Å localization box size).

contribution is insignificant as described by Equation (4.25). However, using the classical definition of J (i.e., $J_T = J_U$) does not account for the entropic contribution which becomes significant at higher temperatures, as underscored by the difference between the total J-integral (J_T) and the energetic J-integral (J_U) computed at 300 K as shown in Figure 4.12. As the temperature increases, the random thermal motion of the carbon atoms also increases, and the entropic contribution becomes more significant, leading to greater than 50% error in the predicted value of total J (J_T), as illustrated in Figure 4.12. The classical J-integral evaluation scheme cannot account for this discrepancy, but the "exact" partition function method (J_T) and the "approximate" cohesive-contour based approach (J_{cohesive}) are able to account for entropic contribution and are in good agreement with each other as can be seen from Figure 4.12. The y-axis on the right-hand side of the plots also depicts the variation in bond order between the atoms just ahead of the crack tip. It can be observed that as the sheet is stretched in tension, the atomic bond gets stretched, and accordingly the bond order decreases as a function of loading (K_I). When the bond order becomes zero the bond breaks, providing a critical value of J_T (J_{Ic}) at crack initiation.

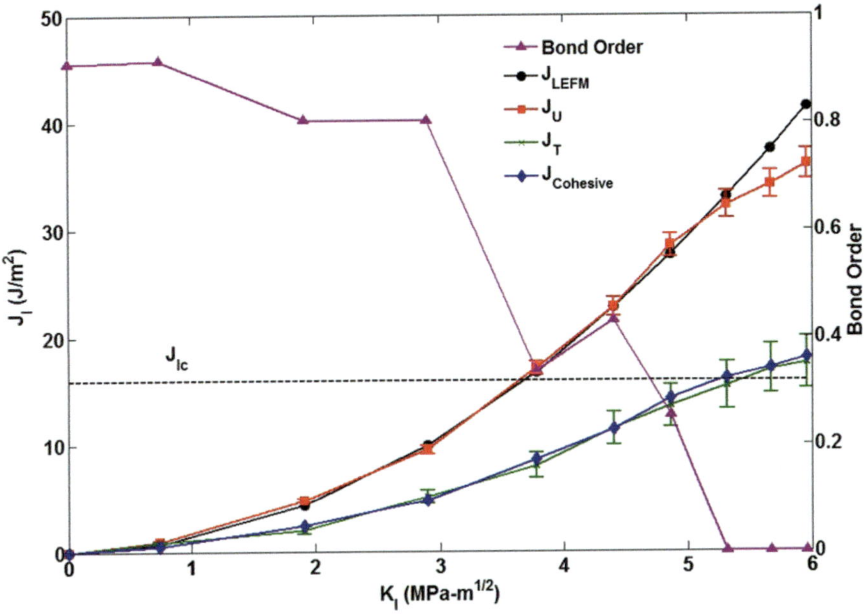

FIGURE 4.12 J-integral computation in graphene as a function of K_{I} using ReaxFF at $T = 300$ K; also shown is change in bond order with loading (27 Å localization box size).

Interestingly, as shown in Figure 4.12, the slopes of the J_T and J_{cohesive} curves change as the bond order goes to zero, and this critical value ($J_{\mathrm{Ic}} = 18\ \mathrm{J/m^2}$) agrees reasonably well with its value published in the literature ($J_{\mathrm{Ic}} = 16\ \mathrm{J/m^2}$, [12]) for nanographene at 300 K. The oscillations in the bond order value at 300 K are likely because of thermal excitation of atoms at elevated temperature. The results indicate that the event of bond breakage corresponds fairly well with when the value of the computed J-integral deviates from LEFM predictions. In addition, the sudden change in the slope of the J_U vs.K_{I} curve in Figure 4.12 (in red) compared with the LEFM (black curve) indicates the point of crack initiation and could directly be related to the fracture toughness of the material system in a predictive sense when bond order data may not be readily available, such as for amorphous materials. This atomistic J-based methodology will be applied for predicting fracture toughness in other materials in later chapters.

4.9 SUMMARY

The J-integral has enjoyed great success as a fracture parameter, especially for materials with nonlinear behavior in the FPZ, and can also be used at bi-material interfaces. However, as described in this chapter, there are significant challenges to implementing the continuum J-integral at the atomistic level due to the discrete (discontinuum) behavior of atoms. In addition, while the continuum J-integral usually ignores entropic effects caused by the random thermal motion of atoms, the atomistic J-integral must include entropic effects based on the fundamental statistical-mechanics definition of free-energy density. A detailed theoretical development of the atomistic J-integral, including entropic effects, as well as results from the application of the atomistic J-integral for a number of case studies at the nanoscale were presented in this chapter. Several case studies involving a center-cracked monolayer graphene sheet at various temperatures were presented. Care should be exercised to ensure that the atomistic J-integral contour encompasses the fully developed fracture process zone for accurate toughness prediction.

REFERENCES

1. Rice, J.R., A path independent integral and the approximate analysis of strain concentration by notches and cracks. *Journal of Applied Mechanics*, 1968. **35**: p. 379–386.
2. Eshelby, J., The elastic energy-momentum tensor. *Journal of Elasticity*, 1975. **5**: p. 321–335.
3. Rigby, R. and M. Aliabadi, Mixed-mode *J*-integral method for analysis of 3D fracture problems using BEM. *Engineering Analysis with Boundary Elements*, 1993. **11**(3): p. 239–256.
4. Jones, R.E., et al., An atomistic *J*-integral at finite temperature based on Hardy estimates of continuumfields. *Journal of Physics: Condensed Matter*, 2010. **23**(1): p. 015002.
5. Jones, R.E. and J.A. Zimmerman, The construction and application of an atomistic *J*-integral via Hardy estimates of continuum fields. *Journal of the Mechanics and Physics of Solids*, 2010. **58**(9): p. 1318–1337.
6. Hardy, R.J., Formulas for determining local properties in molecular-dynamics simulations: Shock waves. *The Journal of Chemical Physics*, 1982. **76**(1): p. 622–628.
7. Weiner, J., *Statistical mechanics of elasticity*. 1983: John Wiley & Sons.
8. Klein, P. and H. Gao, Crack nucleation and growth as strain localization in a virtual-bond continuum. *Engineering Fracture Mechanics*, 1998. **61**(1): p. 21–48.
9. Zhao, P. and G. Shi, Study of Poisson's ratios of graphene and single-walled carbon nanotubes based on an improved molecular structural mechanics model. *Structural Longevity*, 2011. **5**(1): p. 49–58.
10. Van Duin, A.C., et al., ReaxFF: A reactive force field for hydrocarbons. *The Journal of Physical Chemistry A*, 2001. **105**(41): p. 9396–9409.
11. Chenoweth, K., A.C. Van Duin, and W.A. Goddard, ReaxFF reactive force field for molecular dynamics simulations of hydrocarbon oxidation. *The Journal of Physical Chemistry A*, 2008. **112**(5): p. 1040–1053.
12. Zhang, P., et al., Fracture toughness of graphene. *Nature Communications*, 2014. **5**(1): p. 3782.

5 Length-Scale Effects and a Novel *K*-Field Approach to Modeling Fracture

5.1 TRANSITION CRACK LENGTH AND IMPLICATIONS

The theory of flaw size insensitivity was described in Chapter 1. In summary, there exists a critical crack length a_{cr}, which is a material property. Any fracture test with a crack length that is less than the critical crack length $(a < a_{cr})$ results in flaw size insensitivity. The implication of this is that the crack does not propagate in a brittle, self-similar manner. Instead, rupture occurs throughout the material in a uniform manner, despite the crack being present. To visualize the flaw size insensitivity in action, we can consider a crosslinked epoxy polymer system EPON862/DETDA. Referring to Table 5.1, the transition crack length for epoxy resin is ~2590 nm. Typical domain sizes used in molecular dynamics (MD) simulations are considerably smaller due to very high computational cost. Figure 5.1 shows the results from a MD simulation of a SENT fracture test on EPON862 epoxy resin.

The pre-crack length used is 7 nm, which is far below the critical crack length for epoxy, which is 2590 nm. As loading is incremented, the crack does not initiate, even at an applied strain of 15%. Instead, voids appear at various locations throughout the specimen, which is indicative of ductile rupture-based failure, instead of brittle fracture. A detailed analysis of this notch insensitivity effect in epoxy was presented by Sohail and Roy [1].

Flaw size insensitivity is manifested in the form of far-field stress required to fracture a given specimen for a specific material. As the crack length decreases, the far-field fracture stress of the specimen increases $\left(\sigma_{cr,far-field} = \frac{K_{IC}}{\sqrt{\beta \pi a}}\right)$. Since the fracture stress is limited by the theoretical strength of the material, the stress vs. crack length data points deviate from the LEFM equation, and the stress approaches a near-constant value (the theoretical strength). For materials such as silica, the transition crack length is ~1.3 nm as listed in Table 5.1. Park and Kirane [2] have presented an elaborate study of flaw size insensitivity in silica using MD simulations. The far-field fracture stress vs. crack length variation for a SENT test on silica is shown in Figure 5.2.

The blue line in Figure 5.2 is the linear elastic fracture mechanics (LEFM) curve at the experimental fracture toughness value K_{1c}. The theoretical tensile strength of the material, also calculated from MD simulations, is shown as a horizontal dotted line $\sigma_{th} = 12.4$ GPa. The MD data points clearly deviate from the blue curve as the crack length reduces below the critical crack length, approaching σ_{th}. Therefore,

DOI: 10.1201/9781003440734-5

TABLE 5.1
Theoretical Calculations of Critical Semi-crack Length for Notch Insensitivity and Fully Developed Fracture Process Zone Size for Various Materials

Material Type	Room Temperature Fracture Toughness ($MPa\sqrt{m}$)	Room Temperature Theoretical Strength (MPa)	Critical Semi-Crack Length Threshold (nm)	Fully Developed Fracture Process Zone Size (nm)
Monolayer graphene	2.36	108,000	0.15	0.08
Single crystal silicon	1.09	17,000	1.32	3.86
Cross-linked epoxy	0.78	275	2590	11,440
Amorphous carbon	6.00	20,000	28	14

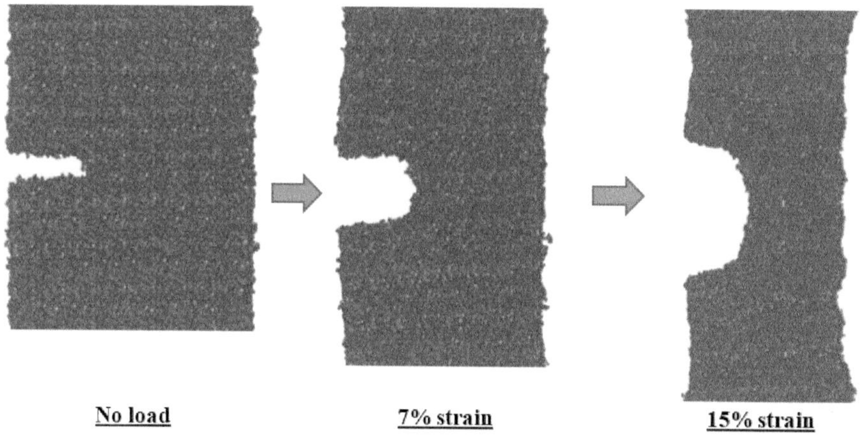

FIGURE 5.1 Edge-notched fracture simulation using molecular dynamics at $T = 300$ K on a 30×30 nm specimen of crosslinked epoxy polymer with a 7 nm pre-crack length.

fracture tests or simulations involving crack lengths below the material's critical crack length ($a < a_{cr}$) will yield incorrect macro-scale fracture toughness predictions, as discussed in Section 1.4 for monolayer graphene. To ensure accurate and verifiable fracture results, a careful consideration of near crack tip fields is required. The subsequent sections in this chapter will build on this concept.

5.2 INTRODUCTION TO ASYMPTOTIC CRACK TIP FIELDS

Based on LEFM, displacement and stress fields near a crack tip can be derived as asymptotic functions. Asymptotic stress fields exhibit a $r^{-1/2}$ dependence, where r is the radial distance from the crack tip. Analytically derived asymptotic field equations

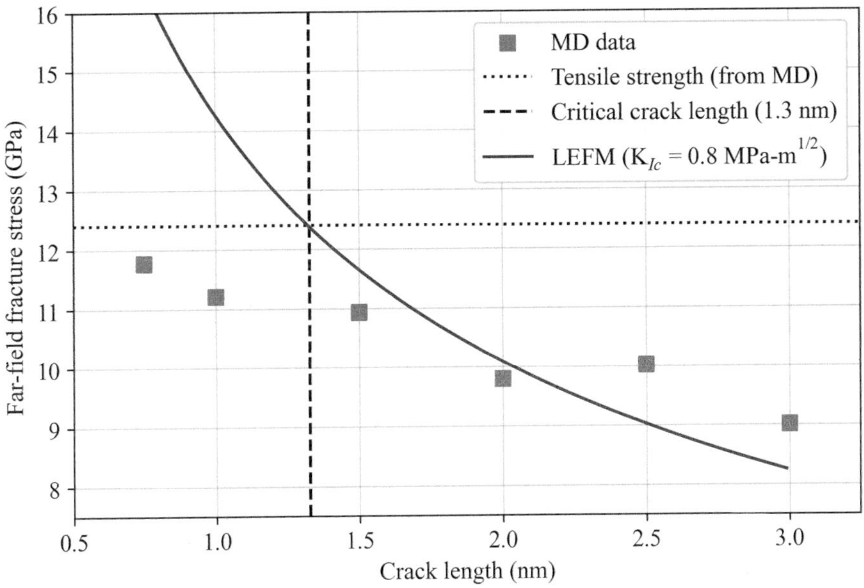

FIGURE 5.2 Far-field fracture stress vs. crack length data from MD fracture simulations on silica using SENT test specimens at $T = 300$ K.

provide a good approximation to the crack tip fields as long as the material behavior is in the linear elastic regime within this zone, known as the zone of K-dominance (D), as defined in Chapter 1. Very close to the crack tip, however, the asymptotic equations predict an unphysical stress singularity as $r \to 0$. In reality, the stresses at the crack tip are finite and are bounded by the theoretical strength of the material. For materials exhibiting small-scale yielding near the crack tip, the LEFM-based asymptotic solution is no longer valid within a radius R from the crack tip, known as the fracture process zone (FPZ). Figure 5.3 graphically shows this deviation for the case of a perfectly plastic material.

In Figure 5.3, the process zone radius is estimated by $R = \frac{1}{2\pi}\left(\frac{K_{1c}}{\sigma_y}\right)^2$ for the PS case, and the radius of K-dominance (D) can be derived using $D = \frac{1}{2\pi}\left(\frac{K_{1c}}{\sigma_0}\right)^2$, where σ_0 and σ_y are the far-field applied stress on a center cracked panel and the yield stress of the material, respectively.

Hence, based on a given yield stress and fracture toughness, a fully developed FPZ size can be calculated, and is listed in Table 5.1 for different materials. Figure 5.4(a) shows a center-cracked panel with equi-biaxial far-field loading. If the material of the panel conforms to small-scale yielding at the crack tip, then the region outside the FPZ (or yield zone) is the zone of K-dominance, where asymptotic displacement and stress fields are applicable, as shown in Figure 5.4(b).

The term "K-dominance" arises from the fact that the field quantities can be expressed as functions of the applied far-field Mode-I, Mode-II, or Mode-III stress intensity factors K_{I}, K_{II}, or K_{III}. Therefore, the effect of individual failure modes can

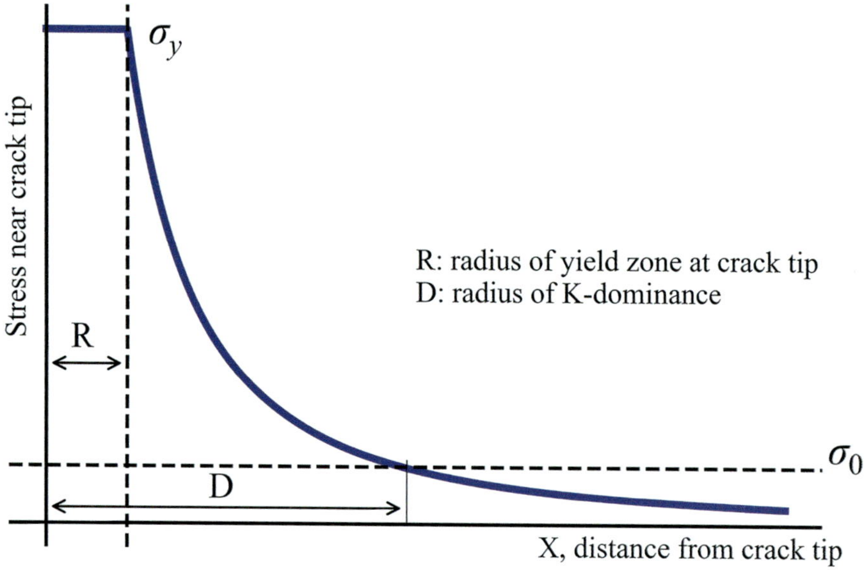

FIGURE 5.3 Limits of asymptotic crack-tip stress fields showing zone of K-dominance.

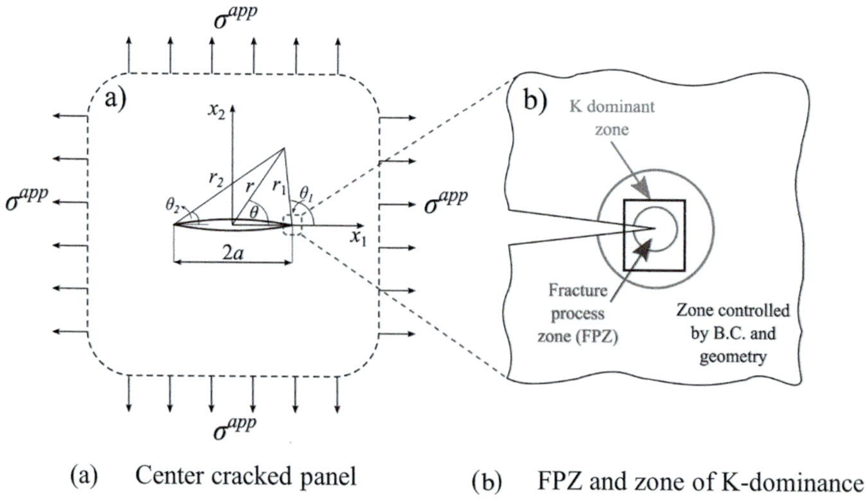

FIGURE 5.4 (a and b) Near crack tip stress zones with equi-biaxial far-field loading on a center-cracked plate. (Published in Andric and Curtin, [3]. https://doi.org/10.1088/1361-651X/aae40c © IOP Publishing. Reproduced with permission. All rights reserved.)

easily be studied by applying LEFM-based displacement boundary conditions to a portion of the simulation domain near the crack with boundaries lying in the zone of K-dominance, as depicted by the black box in Figure 5.4(b). This is particularly useful for the case of fracture at bi-material interfaces, which is inherently a mixed-mode problem, as will be discussed in Section 5.2.2.

5.2.1 Crack Tip Fields in Homogeneous Isotropic Materials

Figure 5.4(a) shows a center-cracked panel with a crack length $2a$, made of a homogeneous isotropic material, that is subjected to far-field stress σ^{app} in both X and Y directions. For isotropic materials, the asymptotic crack tip fields depend on the Young's modulus and Poisson's ratio of the material (or any two elastic constants). A tabulation of all the stress and displacement components under Mode-I, Mode-II, and Mode-III loading is provided in Wang [4]. In this book, we limit our attention to primarily in-plane deformation (Mode-I and Mode-II) as our focus is on solving problems involving plane stress (PS) or plane strain (PE) conditions. The following are the displacement fields as a summation of Mode-I and Mode-II stress intensity factors K_{I} and K_{II}, expressed in polar coordinates relative to the crack tip (r,θ) as depicted in Figure 5.4(a) for a homogeneous isotropic elastic panel.

$$u_1(r,\theta) = \frac{K_{\text{I}}}{2\mu}\sqrt{\frac{r}{2\pi}}\cos\frac{\theta}{2}\left[\kappa - 1 + 2\sin^2\frac{\theta}{2}\right] + \frac{K_{\text{II}}}{2\mu}\sqrt{\frac{r}{2\pi}}\sin\frac{\theta}{2}\left[\kappa + 1 + 2\cos^2\frac{\theta}{2}\right] \tag{5.1a}$$

$$u_2(r,\theta) = \frac{K_{\text{I}}}{2\mu}\sqrt{\frac{r}{2\pi}}\sin\frac{\theta}{2}\left[\kappa + 1 - 2\cos^2\frac{\theta}{2}\right] + \frac{K_{\text{II}}}{2\mu}\sqrt{\frac{r}{2\pi}}\cos\frac{\theta}{2}\left[\kappa - 1 - 2\sin^2\frac{\theta}{2}\right] \tag{5.1b}$$

The above displacement field equations depend on the material's shear modulus (μ) and Poisson's ratio (ν). The constant κ depends on PE or PS constraint.

$$\kappa = 3 - 4\nu \text{ (plane strain)} \tag{5.2a}$$

$$\kappa = (3 - \nu)/(1 + \nu) \text{ (plane stress)} \tag{5.2b}$$

In order to correctly apply the above displacement field boundary conditions to a MD simulation domain, the boundaries must lie in the zone of K-dominance, as schematically depicted by the black box circumscribing the blue circle in Figure 5.4(b). Therefore, an estimation of the size and shape of the FPZ is required. As a first step, it is essential to understand the stress fields. The equations below summarize the stress fields for an isotropic material under combined Mode-I and Mode-II loading.

$$\sigma_{xx}(r,\theta) = \frac{K_{\text{I}}}{\sqrt{2\pi r}}\cos\frac{\theta}{2}\left[1 - \sin\frac{\theta}{2}\sin\frac{3\theta}{2}\right] - \frac{K_{\text{II}}}{\sqrt{2\pi r}}\sin\frac{\theta}{2}\left[2 + \cos\frac{\theta}{2}\cos\frac{3\theta}{2}\right] \tag{5.3a}$$

$$\sigma_{yy}(r,\theta) = \frac{K_I}{\sqrt{2\pi r}} \cos\frac{\theta}{2}\left[1 + \sin\frac{\theta}{2}\sin\frac{3\theta}{2}\right] + \frac{K_{II}}{\sqrt{2\pi r}} \sin\frac{\theta}{2}\cos\frac{\theta}{2}\cos\frac{3\theta}{2} \quad (5.3b)$$

$$\tau_{xy}(r,\theta) = \frac{K_I}{\sqrt{2\pi r}} \cos\frac{\theta}{2}\sin\frac{\theta}{2}\cos\frac{3\theta}{2} + \frac{K_{II}}{\sqrt{2\pi r}} \cos\frac{\theta}{2}\left[1 - \sin\frac{\theta}{2}\sin\frac{3\theta}{2}\right] \quad (5.3c)$$

The out-of-plane component of stress $\sigma_{zz} = \nu(\sigma_{xx} + \sigma_{yy})$ under PE constraint, and $\sigma_{zz} = 0$ for PS. A convenient quantity used to map yielding and the FPZ is the scalar von-Mises stress, which is written in terms of the principal stresses σ_1, σ_2, and σ_3 as follows.

$$\sigma_{vm} = \frac{1}{\sqrt{2}}\left[(\sigma_1 - \sigma_2)^2 + (\sigma_2 - \sigma_3)^2 + (\sigma_3 - \sigma_1)^2\right]^{1/2} \quad (5.4)$$

The general shapes of the FPZ contours based on the stress fields given by Equations (5.3a)–(5.3c) are shown in Figure 5.5.

The Mode-I FPZ contours are shown in blue color and the Mode-II contours are shown in red at a common stress intensity factor, $K_I = K_{II}$. The Mode-II FPZ size is significantly larger than the Mode-I case. Differences in the FPZ due to constraint

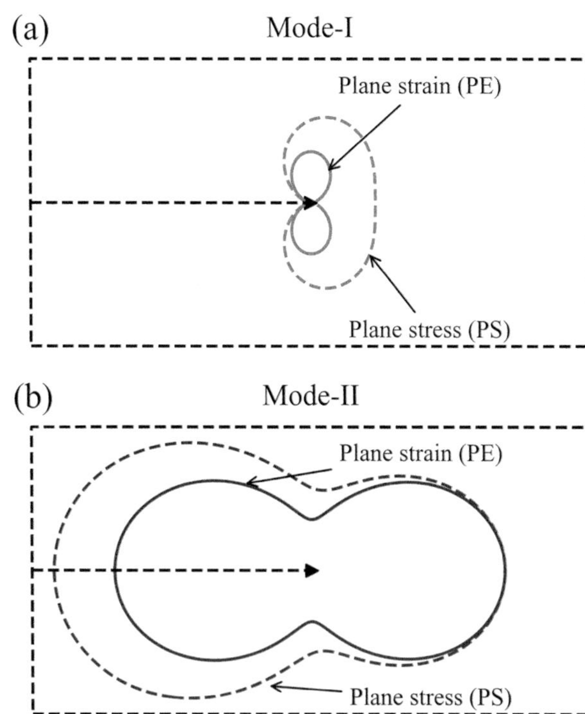

FIGURE 5.5 FPZ in (a) Mode-I PS (blue dashed line) and Mode-I PE (blue solid line), (b) Mode-II PS (red dashed line) and Mode-II PE (red solid line) using a Poisson's ratio of 0.4.

(PS or PE) arise from Poisson's effect, which factors into the σ_{zz} stress component. For the contours shown in Figure 5.5, a Poisson's ratio of 0.4 is used.

In summary, in order to simulate brittle fracture in materials that conform with LEFM, and predict the fracture toughness correctly, the K-test rule must be followed. The K-test rule can be summarized as: radius of FPZ (r_{FPZ}) < domain of K-test < radius of K-dominance $(r_{K\text{-dominance}})$. Hence, boundary conditions for K-test simulations can be applied using Equations (5.1a) and (5.1b). The elegance of the K-test concept is further underscored by the fact that the actual crack length in the K-test domain is automatically assumed to be greater than the critical crack length for notch sensitivity, because notch sensitivity at the crack tip is implicitly prescribed when the K-field is applied.

5.2.2 MD Simulation of Crack Tip Fields at Bi-material Interfaces

Bi-material interfaces arise in problems with material heterogeneity, for example, at the interface between fiber and matrix in fiber-reinforced composites, nanoparticle-polymer interfaces, or at grain boundaries in crystalline microstructures. The analytical stress and displacement field equations pertaining to bi-material interfacial cracks can be found in articles by Suo, Hutchinson, Rice, and Sih [5–8]. Rice and Sih [6] derived the displacement fields using complex analysis and Goursat functions. A summary of all displacement and stress fields in the bi-material problem can be found in [9].

The displacement fields for "Material 1" of the bi-material problem, as shown in Figure 5.6, take the following form, in terms of far-field Mode-I and Mode-II stress intensity factors K_{I} and K_{II}.

$$u_1 = \frac{\sqrt{2\pi r}}{4\pi\mu_1 \cosh(\pi\varepsilon)} \left\{ K_{\text{I}} \left[\kappa_1 \omega_1 h_{11} - \frac{1}{\omega_1} h_{12} + \omega_1 h_{13} \right] + K_{\text{II}} \left[\kappa_1 \omega_1 h_{21} - \frac{1}{\omega_1} h_{22} + \omega_1 h_{23} \right] \right\} \tag{5.5a}$$

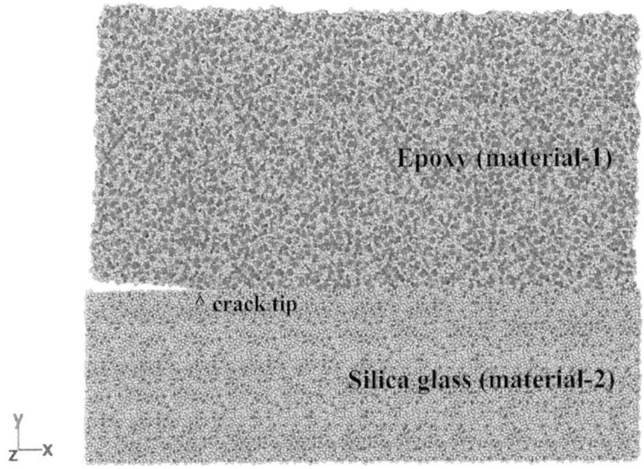

FIGURE 5.6 Bi-material interfacial fracture representation at interface of epoxy and silica glass.

$$u_2 = \frac{\sqrt{2\pi r}}{4\pi\mu_1 \cosh(\pi\varepsilon)} \left\{ K_{\mathrm{I}} \left[\kappa_1 \omega_1 h_{21} - \frac{1}{\omega_1} h_{22} - \omega_1 h_{23} \right] + K_{\mathrm{II}} \left[-\kappa_1 \omega_1 h_{11} + \frac{1}{\omega_1} h_{12} + \omega_1 h_{13} \right] \right\} \tag{5.5b}$$

The variable ε is the bimaterial constant, which is a function of the Poisson's ratios (ν_1 and ν_2) and shear moduli (μ_1 and μ_2) of the two materials, given by the following equation.

$$\varepsilon = \frac{1}{2\pi} \ln \left(\frac{\mu_2 \kappa_1 + \mu_1}{\mu_1 \kappa_2 + \mu_2} \right) \tag{5.6}$$

κ_1 and κ_2 in Equation (5.6) are the constants described in Equation (5.2a) for PE and Equation (5.2b) for PS for the two materials, respectively. The bi-material constant is a key parameter describing the "elastic mismatch" between the two materials. In Equations (5.5a) and (5.5b), h are oscillatory functions, dependent on the angle and radial distance from the crack tip, the details of which can be found in [9]. Similar equations exist for "Material-2" of the bi-material problem, shown in Figure 5.6. The bi-material field equations reduce to the K-field equations for homogeneous materials given by Equations (5.1a), (5.1b), (5.3a)–(5.3c), when the elastic properties of both materials are set to be equal and the bi-material constant ε reduces to zero. The stress fields in 'Material-1' of the bi-material problem are given by the following equations.

$$\sigma_{11} = \frac{1}{2\sqrt{2\pi r} \cosh(\pi\varepsilon)} \left\{ K_{\mathrm{I}} \left[f_{11}^{1} \omega_1 - \frac{1}{\omega_1} \cos(\theta - \theta^*) \right] + K_{\mathrm{II}} \left[f_{11}^{2} \omega_1 - \frac{1}{\omega_1} \sin(\theta - \theta^*) \right] \right\} \tag{5.7a}$$

$$\sigma_{22} = \frac{1}{2\sqrt{2\pi r} \cosh(\pi\varepsilon)} \left\{ K_{\mathrm{I}} \left[f_{22}^{1} \omega_1 + \frac{1}{\omega_1} \cos(\theta - \theta^*) \right] + K_{\mathrm{II}} \left[f_{22}^{2} \omega_1 + \frac{1}{\omega_1} \sin(\theta - \theta^*) \right] \right\} \tag{5.7b}$$

$$\sigma_{12} = \frac{1}{2\sqrt{2\pi r} \cosh(\pi\varepsilon)} \left\{ K_{\mathrm{I}} \left[f_{12}^{1} \omega_1 - \frac{1}{\omega_1} \sin(\theta - \theta^*) \right] + K_{\mathrm{II}} \left[f_{12}^{2} \omega_1 + \frac{1}{\omega_1} \cos(\theta - \theta^*) \right] \right\} \tag{5.7c}$$

Similar to h, f are oscillatory functions relevant to the stress fields, the detailed forms of which can be found in [9]. As an example of the FPZ shapes in the bi-material problem, we consider 'Material-1' to be epoxy resin with a Young's modulus of 5 GPa and Poisson's ratio of 0.4. 'Material-2' is silica glass with a Young's modulus of 100 GPa and Poisson's ratio of 0.25. The yield stress in silica glass is significantly higher than that of epoxy resin, because of which there is minimal yielding that occurs in silica relative to epoxy. However, for purposes of illustration, the yield stress for both materials is taken to be equal. The FPZ based on the bi-material K-field Equations (5.7a)–(5.7c) and the von-Mises stress (Equation (5.4)) is shown in Figure 5.7.

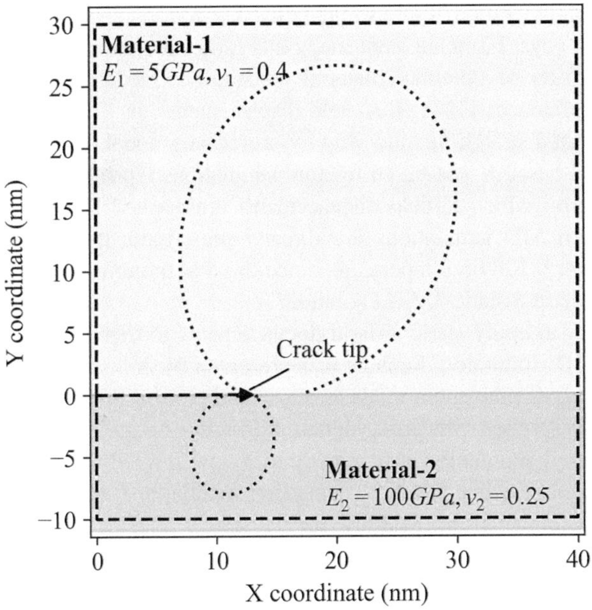

FIGURE 5.7 Representative FPZ contour based on von-Mises stress calculated using bi-material K-field equations under Mode-I, PE loading.

PE conditions are used in the representative FPZ contour plot. Comparing the Mode-I, PE contour for the homogeneous material case (Figure 5.5a) with the bi-material contour reveals the effect of elastic mismatch on the near crack tip stress distribution. As can be seen from the angled orientation of the FPZ contours in Figure 5.6, far-field pure Mode-I loading can give rise to mixed-mode deformation and stresses at the crack tip due to material property mismatch.

5.3 INTRODUCTION TO THE *K*-TEST SIMULATION FRAMEWORK

In the previous section, the asymptotic stress and displacement fields around a crack tip, known as K-fields, were described. The fields are a function of the far-field stress intensity factor (K_I, K_{II}, or K_{III}) and depend on the elastic material properties, mode mix, and constraint (PE or PS). The cases of homogeneous isotropic materials and bi-material interfaces were described, which have important applications in fracture scenarios. In this section, we describe how the K-fields can be used to accurately perform fracture simulations near the crack tip.

The zone of K-dominance is shown in Figure 5.4(b), where the K-field equations described in Section 5.2 are applicable. The "K-test" is a novel simulation technique in which the fracture simulation domain (black box drawn in Figure 5.4(b)) is chosen to be a subset of the full fracture test (e.g. plate under equi-biaxial loading shown in Figure 5.4(a)). By choosing the reduced simulation domain, the K-field displacements or stresses may be applied as boundary conditions, as opposed to applying far-field load directly, which would unnecessarily increase the MD domain size. In the literature on near crack-tip fracture simulations, the "K-test" approach has been used in a few different ways. Finite element method (FEM) simulations can be conducted to explore the effects of complex material constitutive models on the FPZ as was performed in references [10–12]. K-field displacements, in FEM K-test simulations, can be applied in a straightforward way by prescribing nodal displacement boundary conditions. MD-based K-test-based fracture simulations typically involve a boundary layer of atoms on which K-field displacements are applied [13, 14]. If the K-field displacements in MD simulations are directly prescribed, the simulation typically needs to be at zero Kelvin temperature, since thermal motion of the boundary atoms is not admissible in a static K-field solution.

A novel way to apply static K-field displacements to thermally vibrating boundary atoms in MD simulation domains was presented by Aditya et al. [15]. Instead of directly applying displacement fields to individual atoms, atoms can be tethered and displaced using springs with pre-assigned stiffness. The spring-based method, also known as steered molecular dynamics (SMD), is derived from the anchor-point approach used in certain multi-scale coupling simulations, as will be described in Chapter 6. The tethered springs allow for physically realistic thermal movement of the boundary atoms while keeping the displacement at the other end of the spring fixed as mandated by the K-test boundary condition, thereby allowing SMD simulations to be run at finite temperatures.

5.3.1 Example of Atomistic Fracture Simulation of Epoxy

The K-test simulation framework with spring-based boundary loading enables accurate fracture simulations at the nanoscale and at finite temperature. In this section, an example of the application of K-test to Mode-I fracture simulations on epoxy polymer is presented. The K-test geometry and applied K-field displacements for epoxy resin are shown in Figure 5.8.

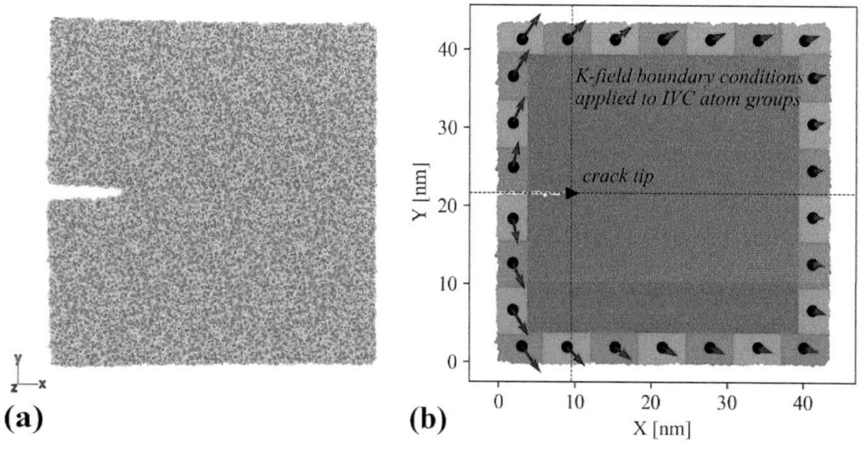

FIGURE 5.8 (a) Atomistic K-test simulation domain with pre-crack, (b) visualization of displacement vectors applied to MD domain boundary atom groups.

TABLE 5.2
Material Properties for Crosslinked Epoxy Resin Computed from Uniaxial Tension Tests Using MD Simulations

Temperature	Young's Modulus (GPa)	Poisson's Ratio	0.2% Offset Yield Stress (MPa)	Max. Stress Prior to Softening (MPa)
$T = 1$ K	5.2	0.35	169	314
$T = 300$ K	2.6	0.35	95	184

The Mode-I K-test is controlled by the far-field stress intensity factor (K_I) and depends on the Young's modulus and Poisson's ratio (refer to Equations (5.1a) and (5.1b)). The elastic properties for the epoxy resin are first computed using separate uniaxial tension MD simulations. The estimated material properties are shown in Table 5.2 at temperatures of $T = 300$ K (room temperature) and $T = 1$ K, which is useful when studying fracture results by eliminating noise from thermal fluctuations. Temperature strongly affects the elastic and yield properties of materials as can be seen in Table 5.2, which in turn impacts the K-test simulation.

The MD K-tests performed here are at finite temperatures, i.e., atomic motion is simulated via time integration (Verlet algorithm). Therefore, simulations include inertial effects due to the finite mass of atoms. The K-test loading is applied in increments of $K_\mathrm{I} = 0.01$ MPa$\sqrt{\mathrm{m}}$ by prescribing K-field displacements via springs to the boundary atom groups, each of which is shown in Figure 5.8(b) as orange and green sub-cells. Altogether, 26 sub-cells are employed, the centroid of each sub-cell being attached to a spring for SMD. Each loading increment takes place over a finite amount of simulation time, firstly ramping up the boundary atom displacements to the prescribed positions, and secondly, allowing equilibration of the overall simulation domain to take place. The exact time over which each loading step must take place depends on the specific materials, geometry, and loading conditions. The potential energy of the entire MD system is used, as shown in Figure 5.9, to study whether equilibration of the system has taken place for a given load step, before proceeding to the next load step, so as to prevent error propagation. Each load step takes place over 80 ps of simulation time, during which the potential energy rises until the boundary atoms have reached their prescribed positions, followed by gradual equilibration as the load is redistributed through the entire system via bond relaxation.

A powerful feature of the K-test framework is that validation of the applied loading can be performed by studying the stress fields inside the simulation domain due to the applied displacement fields at the simulation boundaries. For example, we can plot the stress along a line ahead of the crack tip and compare with the available K-field stress solution (i.e., setting $\theta = 0$ in Equations (5.1a) and (5.1b)). The von-Mises stress from both MD simulation results and from the K-field (LEFM curve) are compared, shown in Figure 5.10. The MD results are in excellent agreement with the LEFM curve, deviating only when the material yield stress is exceeded, i.e., when inside the FPZ. Inside the yield zone, the stress reaches a plateau as is predicted based on Figure 5.3.

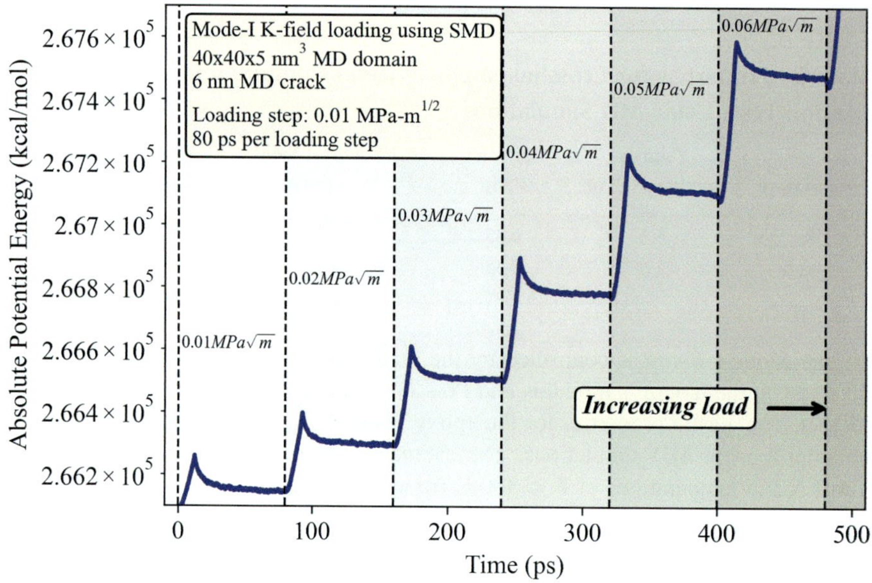

FIGURE 5.9 MD K-test system potential energy convergence over load steps for epoxy resin Mode-I fracture simulation at $T = 1$ K.

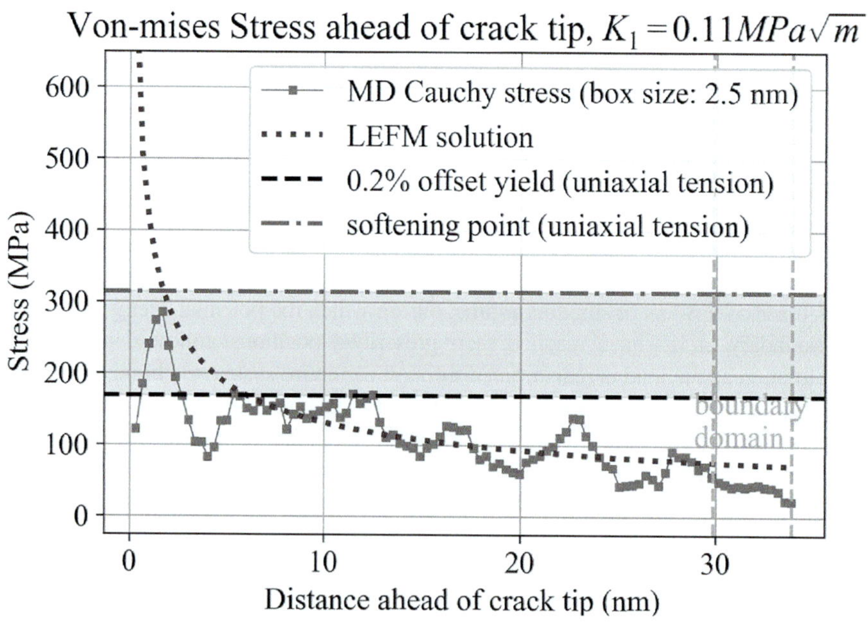

FIGURE 5.10 Comparison of MD-computed stress ahead of the crack tip and the K-field (LEFM) analytical curve for epoxy resin K-test at $K_1 = 0.01$ MPa$\sqrt{\text{m}}$.

5.3.2 Correspondence Between Stress Relaxation and Void Nucleation in the FPZ in Epoxy

In this section, we demonstrate how the present K-field-based MD technique can simulate void nucleation and void growth in the FPZ near a crack tip, which are difficult to simulate using purely continuum techniques, such as FEM, because void location and growth law have to be defined *a priori*. On the other hand, in MD, the void growth occurs automatically based on the bond energy potential used and the distribution of the free volume of the polymer. MD simulation results for von-Mises stress and strain for the PE case at $T = 1$ K, loaded up to $K_1 = 0.39$ MPa$\sqrt{m}$, are shown in Figure 5.11.

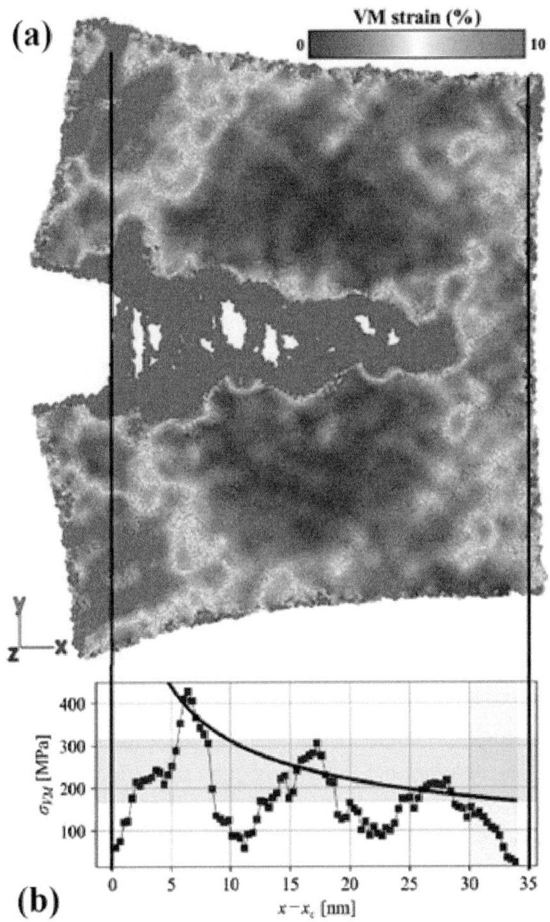

FIGURE 5.11 (a) von-Mises strain contour computed using MD (b) corresponding MD computed von-Mises stress in epoxy plotted as a function of distance ahead of crack tip for PE case at $T = 1$ K and $K_1 = 0.39$ MPa$\sqrt{m}$. Solid black curve represents von-Mises stress obtained from LEFM.

In Figure 5.11(a) the von-Mises strain in the MD domain is visualized as a contour plot using the OVITO software [16]. The corresponding von-Mises stress ahead of the crack tip is shown in Figure 5.11(b). Figure 5.11(a) clearly shows, along with the strain contour, the formation of voids and fibrils due to stretched polymer chains, which is a common feature of failure in glassy polymers [17]. By comparing the von-Mises strain contour with the stress plot, the occurrence of voids is found to coincide with zones of reduced von-Mises stress, as is expected. In between the zones of reduced stress are regions of high stress, which coincide with unbroken ligaments containing stretched fibrils of epoxy molecules, as visible in Figure 5.11(a). These fibrils are known to bear significantly higher load than the bulk polymer as the molecules in the fibrils are fully stretched and dihedral angle transitions in the epoxy molecular structure have been fully exhausted. The strain levels in the fibrils exceed 10% as shown in Figure 5.11(a). It is notable that the local von-Mises stress peaks in Figure 5.11(b) corresponding to the fibril locations in Figure 5.11(a), and exceed the horizontal green band of yield stress for epoxy as shown in Figure 5.11(b), corroborating the previous statement. At $K_1 = 0.39\ \text{MPa}\sqrt{\text{m}}$, the LEFM von-Mises stress (plotted as a solid black line) is fully above the yield stress, i.e., theoretically the entire MD domain is expected to have yielded at this point and K-test conditions are no longer satisfied. Significant deviation of stresses from LEFM occurs well before this point and, as was explained in previous sections, the validity of K-test results becomes dubious once significant inelastic yielding has taken place throughout the MD simulation domain. Consequently, the actual fracture toughness of epoxy ($K_{1c} = 0.78\ \text{MPa}\sqrt{\text{m}}$) cannot be predicted using the K-field approach unless the K-field can be placed at least at a radius of 11,440 nm from the crack tip, which is the FPZ size for epoxy as tabulated in Table 5.1. Of course, this is not feasible with the computational resources currently available. Despite this drawback, the von-Mises stress data shown in Figure 5.11(b) from MD simulation appear to match reasonably well with the LEFM curve shown as a solid black line, with abrupt deviations due to voids and fibril formation, exactly as one might expect to occur within a FPZ.

5.3.3 Case Study of Atomistic Fracture Simulation at a Bi-material Interface

As a case study of bi-material interfacial fracture, the epoxy/silica interfacial problem discussed previously in Section 5.2.2 is analyzed using MD simulation. A polymer block consisting of 85% crosslinked epoxy resin (EPON862/DETDA) is modeled bonded to a substrate of silica glass, which is representative of fiber/matrix interfaces in glass fiber-reinforced polymer composite materials. There are no covalent bonds at the interface, and only non-bonded interactions are allowed at the epoxy/silica interface. Based on uniaxial tension tests using MD, the elastic properties of the epoxy (Material-1) and silica glass (Material-2) are $E_1 = 5$ GPa, $\nu_1 = 0.4$, $E_2 = 100$ GPa, and $\nu_2 = 0.25$. These properties are used to calculate the bi-material K-dominant

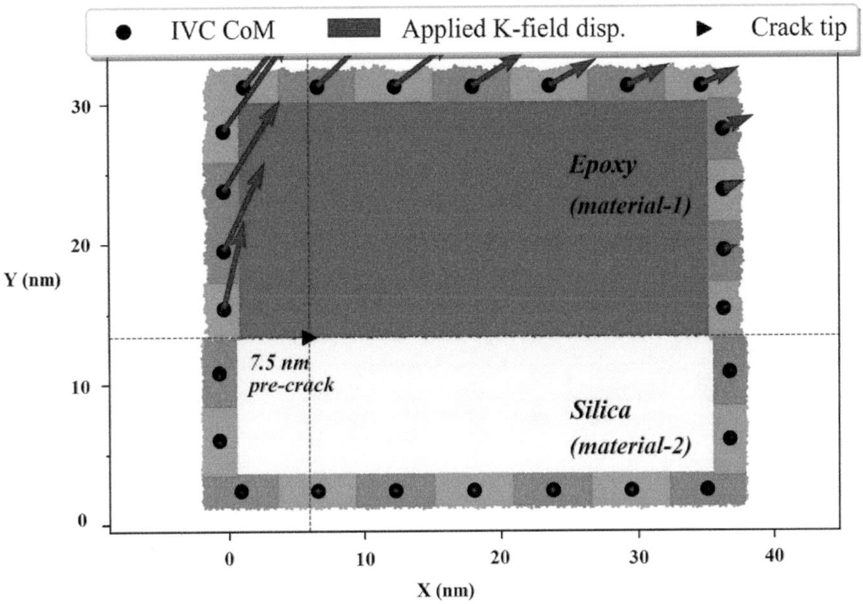

FIGURE 5.12 Schematic of epoxy/silica system for interfacial fracture simulations using K-field displacements applied at boundaries via steered MD using springs.

displacement fields, which are used to apply the appropriate near crack tip deformations to the bi-material system, as shown in Figure 5.12.

Fracture simulations in this case study are performed at a low temperature $T = 1$ K so that thermal fluctuations in stresses can be minimized. This is important when studying all-atom MD systems of polymeric materials having a high number of degrees of freedom per unit volume, which can make virial stress calculations subject to thermal "noise" at elevated temperatures.

For simulating fracture at a epoxy/silica bi-material interface, Mode-I loading under PE conditions is applied in increments of $\Delta K_{\mathrm{I}} = 0.01\ \mathrm{MPa}\sqrt{\mathrm{m}}$ on the K-field domain boundary as shown by the green and orange cells in Figure 5.12 based on Equations (5.5a) and (5.5b). As can be seen in Figure 5.12, the applied K-field boundary displacements on epoxy shown by the blue arrows are much larger than the ones on silica glass due to the large mismatch in their Young's modulus.

Based on MD simulations, crack initiation in the bi-material system occurs at approximately $K_{\mathrm{I}} = 0.04\ \mathrm{MPa}\sqrt{\mathrm{m}}$. To validate the K-field-based steered MD method for bi-material systems, the MD computed normal stress (σ_{yy}) ahead of the crack tip is plotted along the bi-material interface and compared with LEFM predictions (Equation (5.7b)). The stress vs. distance (from crack tip) plot is shown

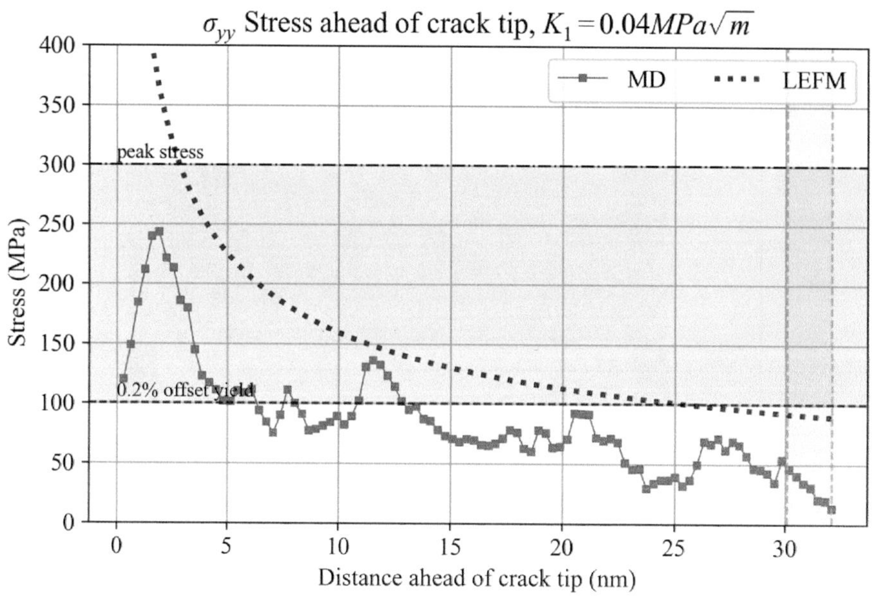

FIGURE 5.13 σ_{yy} stress component plotted vs. distance from crack tip along the interfacial line ahead of the crack tip. MD data points and the LEFM curve are compared at a far-field loading of $K_I = 0.04$ MPa$\sqrt{\text{m}}$.

in Figure 5.13 at $K_I = 0.04$ MPa$\sqrt{\text{m}}$. MD stresses are computed using volume averaging boxes of size 2.5 nm in each direction in the epoxy.

In Figure 5.13, the green region is bounded by the 0.2% yield stress of the epoxy (lower limit) and the peak stress of the epoxy prior to softening (upper limit). The vertical orange region in Figure 5.13 refers to the K-test boundary domain upon which steered MD loading is applied. The stress data in the orange boundary region is deemed not accurate due to boundary effect.

The MD-predicted stress data in Figure 5.13 shows good agreement with the LEFM trend, though slightly conservative. The underprediction is likely due to the volume averaging involved in the MD stress calculations which tends to result in reduced stress concentration. Qualitatively, the data points closer to the crack tip exhibit yielding behavior and likely void and fibril formation in the epoxy at the interface, inferred from the stress peak and trough between 0 and 12 nm distance from the crack tip. The crack initiation and growth at the epoxy/silica bi-material interface are visualized using MD in Figure 5.14 at various K_I loading levels. Crack growth is preceded by the formation and coalescence of voids ahead of the crack tip. Eventually, the crack grows all the way through the interface, with some detached epoxy molecules adhering to the silica surface as shown in Figure 5.14.

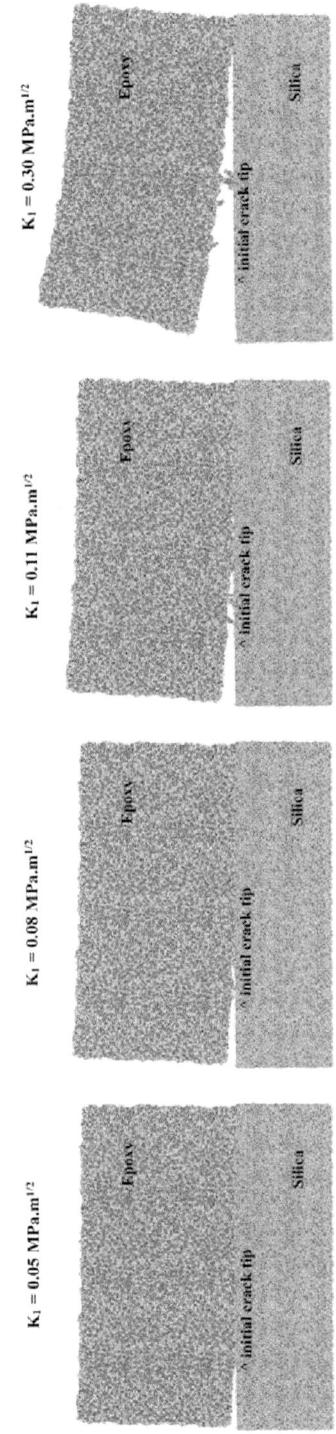

FIGURE 5.14 MD visualization of crack initiation and growth at epoxy/silica bi-material interface at various K_I loading levels.

5.4 SUMMARY

In this chapter, essential practical aspects pertaining to correctly modeling fracture, especially at the microscale, were delineated. This includes the concept of flaw size insensitivity, which results in a transition crack length below which brittle fracture does not occur. This was demonstrated for epoxy resins using MD simulations. The important concept of K-field was introduced, for homogeneous materials and for bi-material systems. Finally, the K-test simulation framework was described, and examples were shown for fracture simulation in epoxy and epoxy/silica bi-material interface.

The primary conclusion from this chapter is that in order to simulate brittle fracture in materials that conform with LEFM, and to predict the fracture toughness correctly, the K-test rule must be followed. The K-test rule can be summarized as: radius of FPZ (r_{FPZ}) $<$ domain of K-test boundary $<$ radius of K-dominance ($r_{K\text{-dominance}}$). Hence, boundary conditions for K-test simulations must be applied in a region fully encompassing the FPZ but contained within the zone of K-dominance, using Equations (5.1a) and (5.1b). The elegance of the K-test concept is further underscored by the fact that the actual crack length in the K-test domain is automatically assumed to be greater than the critical crack length for notch sensitivity, because notch sensitivity at the crack tip is implicitly prescribed when the K-field is applied.

The MD-based K-test approach was validated for simulating crack growth in monolithic epoxy block and for epoxy/silica bi-material system. This methodology is applicable to materials with small-scale yielding and will not be accurate for materials where the FPZ is larger than the zone of K-dominance, such as ductile materials exhibiting large-scale plasticity, as that would violate the K-test rule.

REFERENCES

1. Sohail, T. and S. Roy, An investigation of the validity of LEFM at the nanoscale in amorphous materials using the atomistic J-integral including entropic effect. *Theoretical and Applied Fracture Mechanics*, 2023. **125**: p. 103913.
2. Park, J. and K. Kirane, Transitional flaw size sensitivity of amorphous silica nanostructures analyzed by ReaxFF/SiO based molecular dynamics. *Journal of Applied Physics*, 2021. **129**(17): p. 175103.
3. Andric, P. and W.A. Curtin, Atomistic modeling of fracture. *Modelling and Simulation in Materials Science and Engineering*, 2018. **27**(1): p. 013001.
4. Wang, C.H., *Introduction to fracture mechanics*. 1996: DSTO Aeronautical and Maritime Research Laboratory Melbourne, Australia.
5. Hutchinson, J.W. and Z. Suo, Mixed mode cracking in layered materials. *Advances in Applied Mechanics*, 1991. **29**: p. 63–191.
6. Rice, J. and G.C. Sih, Plane problems of cracks in dissimilar media. *Journal of Applied Mechanics*, 1965. **32**(2): p. 418–423.
7. Hutchinson, J.W., M. Mear, and J.R. Rice, Crack paralleling an interface between dissimilar materials. *Journal of Applied Mechanics*, 1987. **54**(4): p. 828–832.
8. Rice, J., Elastic fracture mechanics concepts for interfacial cracks. *Journal of Applied Mechanics*, 1988. **55**: p. 98–103.

9. Gu, Y., J. Lin, and F. Wang, Fracture mechanics analysis of bimaterial interface cracks using an enriched method of fundamental solutions: Theory and MATLAB code. *Theoretical and Applied Fracture Mechanics*, 2021. **116**: p. 103078.
10. Tvergaard, V. and J.W. Hutchinson, The relation between crack growth resistance and fracture process parameters in elastic-plastic solids. *Journal of the Mechanics and Physics of Solids*, 1992. **40**(6): p. 1377–1397.
11. Lai, J. and E. Van der Giessen, A numerical study of crack-tip plasticity in glassy polymers. *Mechanics of Materials*, 1997. **25**(3): p. 183–197.
12. Khan, D., S. Singh, and A. Needleman, Finite deformation analysis of crack tip fields in plastically compressible hardening–softening–hardening solids. *Acta Mechanica Sinica*, 2017. **33**: p. 148–158.
13. Rimsza, J.M., R.E. Jones, and L.J. Criscenti, Crack propagation in silica from reactive classical molecular dynamics simulations. *Journal of the American Ceramic Society*, 2018. **101**(4): p. 1488–1499.
14. Khosrownejad, S.M., J.R. Kermode, and L. Pastewka, Quantitative prediction of the fracture toughness of amorphous carbon from atomic-scale simulations. *Physical Review Materials*, 2021. **5**(2): p. 023602.
15. Aditya, S.S., M.D.A. Amin, and S. Roy, A steered molecular dynamics study of effects of constraint and temperature on fracture process zone evolution in crosslinked epoxy resin. *Mechanics of Advanced Materials and Structures*, 2025: p. 1–15. https://doi.org/10.1080/15376494.2025.2482795
16. Stukowski, A., Visualization and analysis of atomistic simulation data with OVITO–the Open Visualization Tool. *Modelling and Simulation in Materials Science and Engineering*, 2009. **18**(1): p. 015012.
17. Rottler, J., Fracture in glassy polymers: A molecular modeling perspective. *Journal of Physics: Condensed Matter*, 2009. **21**(46): p. 463101.

6 Concurrent Multi-scale Coupling of FEM and MD to Simulate Fracture in Brittle Materials

6.1 INTRODUCTION

In the literature on concurrent coupling, numerous methods have been explored, major ones being the quasi-continuum method [1], the bridging domain method [2], and the coupled atomistic and discrete dislocations method [3]. A comprehensive list of various techniques and the specific choices that need to be made when selecting or developing a concurrent coupling method for a given problem can be found in the review paper by Miller et al. [4]. These considerations include how the domain partitioning takes place (overlapping domains as shown in Figure 6.1 or contiguous), how the MD and FEM domains are constrained to move together, and whether the formulation is energy- or force-based. Some of the methods are designed specifically with crystalline materials in mind, while others may be based on a specific form of MD force field. The objective of this chapter is to present the formulation and specific implementations details for a general concurrent coupling method that can be used for a variety of materials, MD force fields, and make use of existing specialized MD solvers. While the focus of this chapter is on MD-FEM coupling, the approach can be extended to the coupling of MD and any continuum-scale solver, such as peridynamics or the material point method.

6.2 FORMULATION OF CONCURRENT COUPLING METHOD

The concurrent coupling method that is described in this chapter attempts to build on several existing techniques in the literature. One important feature of the method is that there is a bridging domain where the MD and FEM domains overlap, and information can be exchanged, as shown by the overlap region between MD and FEM in Figure 6.1. The use of a bridging domain allows for "weakened compatibility" in terms of the boundary conditions at the domain interface, therefore making the method more flexible in terms of materials that can be modeled and MD force fields that can be used. Using the bridging domain method also eliminates any mesh dependency related to the FEM domain, as opposed to certain methods in which the FEM mesh needs to be refined to the scale of atoms or match with the nanoscale crystal structure. Another important feature of the coupling method is that displacement field constraints between the MD and FEM domains are maintained using

DOI: 10.1201/9781003440734-6

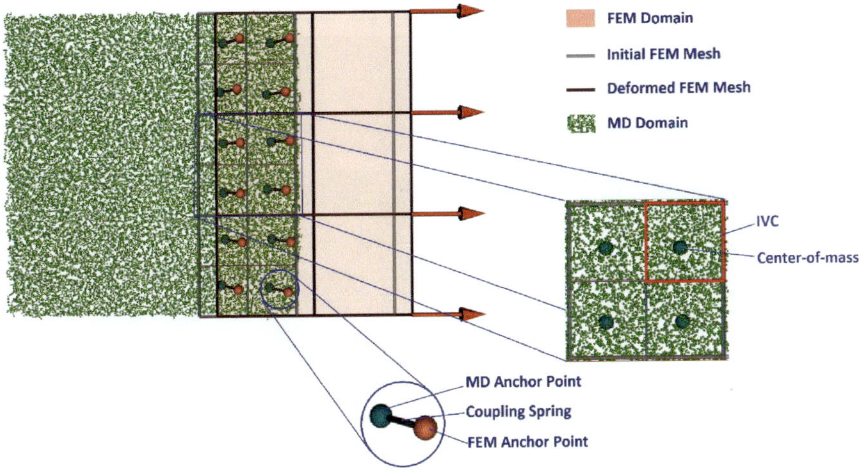

FIGURE 6.1 Schematic of MD-FEM concurrent coupling method with a bridging domain, anchor point-based force transfer, and statistical averaging via interface volume cells (IVC).

springs and anchor points, similar to the Capriccio method presented by Pfaller et al. [5, 6]. Anchor points refer to the two endpoints of a harmonic spring, connecting FEM and MD domains, respectively, in the bridging domain as depicted in Figure 6.1. The use of springs results in a stable force-based method with a tunable spring constant parameter. The spring constant acts as a penalty term with higher values indicating more rigid or "direct" coupling. Finally, the concept of interface volume cells (IVC) [7], as shown in Figure 6.1, is used to ensure appropriate statistical (spatial and temporal) averaging of the discrete MD data, such as atomic displacements, is performed, which is then communicated to the FEM domain. The following subsections describe these fundamental concepts in detail.

6.2.1 Energy Blending Across Bridging Domain

For the bridging domain method to be accurate, it must be recognized that there is a double counting of system energy in the MD-FEM overlap region. In effect, this can make the bridging domain twice as stiff as the remaining domain, which is spurious. Therefore, energy blending must be performed which must follow a partition-of-unity. A 2-D representation of the energy blending concept is shown in Figure 6.2 using an energy weight factor α which has a value of 1 inside the MD domain and 0 inside the FEM domain, with intermediate values in the bridging domain. The representation in Figure 6.2 shows a linear transition of the weight factor α in the MD and FEM domains, though other more complex blending functions can be chosen that still obey partition-of-unity.

The energy weight factor α and the complementary factor $1 - \alpha$ can, in theory, be multiplied by the energy contributions from the MD and FEM domains respectively in a global energy-based formulation. For purely energy-based methods such as the Arlequin method [8] and the original bridging domain method [2], applying the

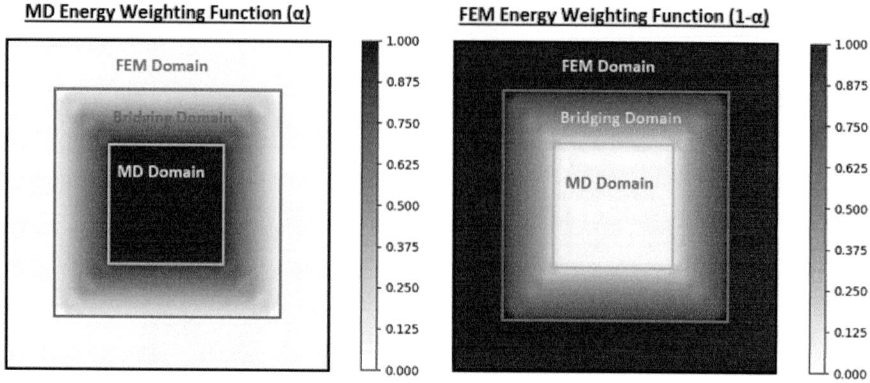

FIGURE 6.2 Representation of energy blending across MD-FEM bridging domain using a weighting factor α.

weighting factors to MD and FEM energy contributions can be performed in a straightforward manner. However, the aforementioned methods have the deficiency that they are based on molecular statics (as opposed to dynamics) and both domains are solved in a single step using a monolithic solver. The method described in this chapter does not have these restrictions and specialized MD solvers such as LAMMPS [9] may be used to run finite temperature MD simulations. However, the disadvantage of this approach is that the application of an energy weight factor to the MD domain is no longer a trivial problem. This issue was recognized in the articles by Pfaller et al. [5, 6] and a workaround was proposed. Due to edge effects in the MD domain (early termination of polymer chains or dangling bonds), it was found that there is a significant reduction in material stiffness in the bridging domain. This reduced stiffness contributes to reducing MD energy in the bridging domain, and hence, precludes the need for an external energy weighting scheme using α.

Energy weighting in the FEM domain can be performed by modifying the stiffness matrix $\left(\boldsymbol{K}^{\mathrm{FE}}\right)$ to include the complementary energy weight factor $1-\alpha$ as follows, with the matrices B and C being the finite element strain transformation matrix and material constitutive matrix respectively, as described in Chapter 2:

$$\boldsymbol{K}^{\mathrm{FE}} = \sum_{e=1}^{n^e} \int_{\Omega_e} (1-\alpha)\, B^{\mathrm{T}} C\, B\, d\Omega_e \tag{6.1}$$

It should be noted that α is a function of position, i.e., $\alpha = \alpha(\boldsymbol{X})$. Therefore, the integral in Equation (6.1), for each element, can be evaluated by considering the interpolated α value at the Gaussian integration points.

6.2.2 Spring-Based Constraint of MD and FEM Displacement Fields

As mentioned in Section 6.2, the spring-based anchor point coupling scheme is an important feature of the present coupling method and was first introduced via the

Capriccio method [5, 6]. The use of springs closely resembles the penalty method of applying constraints to optimization problems. By representing the problem as a global (MD + FEM) energy minimization problem, the Lagrangian (L) for the coupled system can be written as follows:

$$L = T^{\mathrm{MD}} - V^{\mathrm{FE}} - V^{\mathrm{MD}} - V^{\mathrm{pen}} \tag{6.2}$$

In the above, T^{MD} is the kinetic energy in the MD domain, V^{FE} and V^{MD} are the potential energies in the FEM and MD domains respectively, and V^{pen} is the penalty term resulting from the coupling springs. For the case of linear springs, the penalty term can be expressed as follows:

$$V^{\mathrm{pen}} = \frac{1}{2}\left(\boldsymbol{u}^{\mathrm{AP/FE}} - \boldsymbol{u}^{\mathrm{AP/MD}}\right)^{\mathrm{T}} \boldsymbol{K}^{S}\left(\boldsymbol{u}^{\mathrm{AP/FE}} - \boldsymbol{u}^{\mathrm{AP/MD}}\right) \tag{6.3}$$

In the above, $\boldsymbol{u}^{\mathrm{AP/FE}}$ and $\boldsymbol{u}^{\mathrm{AP/MD}}$ are the vectors of the anchor point positions on the FEM side and MD side respectively, as shown in Figure 6.1. Anchor points refer to the two endpoints of a harmonic spring, connecting FEM and MD domains, respectively, in the bridging domain. Anchor points can be defined in several ways depending upon the requirements of the problems.

On the MD side of the bridging domain, anchor points can be defined at either an individual atom or at the center of mass of a group of atoms. In the original Capriccio method [5, 6], a coarse-grained MD representation of a thermoplastic polymer was considered, wherein each particle is a monomer. In their case, anchor points were defined as probabilistically selected individual coarse-grained particles. In Aditya et al. [10], an all-atom MD representation of a thermosetting polymer (epoxy) was considered. Due to the heterogeneity of atoms in the ReaxFF all-atoms system, selecting individual particles as anchor points was not a suitable strategy. Therefore, the atoms in the bridging domain were spatially classified into atomic groups known as IVC [7], and the center-of-mass of each IVC was taken to be an MD anchor point, as shown in Figure 6.1. The strategy of using IVCs also worked well for crystalline materials, for example, graphene, and appears to be a more intuitive choice compared with using individual atoms as anchor points. In all cases, whether defining anchors as individual atoms or the center-of-mass of groups of atoms, atomistic position data in MD simulations is noisy due to the presence of thermal oscillations at finite temperatures. Therefore, time averaging needs to be performed prior to communicating MD anchor point positions to the continuum FEM domain. An example of how time-averaging window size affects the center-of-mass positions of IVC atomic groups in the MD bridging domain is shown in Figure 6.3.

The results presented pertain to coupled simulations on monolayer graphene at $T = 300$ K (room temperature). As the time averaging window size is increased, the oscillations in the displacement of a representative IVC gradually decrease, converging to a single value which can be provided to the coupled FEM solver. As MD-FEM iterations proceed, the IVC displacement progressively changes as shown in Figure 6.3. In this particular case, a 5 ps time averaging window works well to help reduce noise in the thermal oscillations of the IVC center-of-mass.

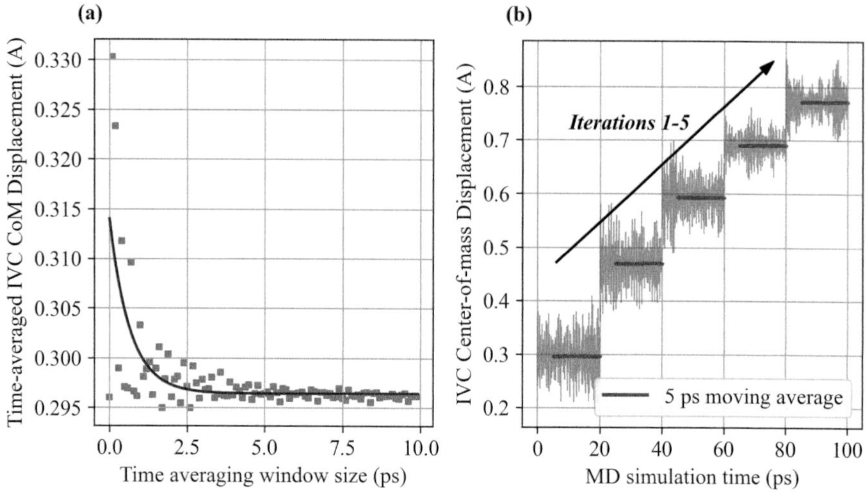

FIGURE 6.3 Results from coupled simulations of monolayer graphene sheets at $T = 300$ K. (a) Effect of time-averaging window size on filtering noise due to thermal oscillations in center-of-mass position data from MD domain IVC atom groups, (b) displacement convergence of IVC center-of-mass vs. time for each load step.

The anchor points on the FEM side can also be defined in a number of ways. Two methods are described here, the first one is based on Lagrange multipliers as used in the Arlequin method [8] and the Capriccio method [5, 6]. The second method is a more direct point-based approach developed by the authors. In both approaches, the MD anchor points are first defined, and then based on their positions, FEM anchor points are duplicated. The difference in the two approaches is regarding how the FEM anchor points are constrained to move with the FEM domain. Using the Lagrange multiplier approach, an FEM anchor point displacement field is generated using the moving least squares (MLS) method. This field is constrained to the FEM nodal displacement fields using the method of Lagrange multipliers as follows, where $\mathbf{\Lambda}$, $\boldsymbol{U}^{\mathrm{AP/FE}}$, and $\boldsymbol{U}^{\mathrm{FE}}$ are field variables for the Lagrange multipliers, FEM anchor point displacements, and FEM nodal displacements respectively, $\boldsymbol{\lambda}$, $\boldsymbol{u}^{\mathrm{AP/FE}}$, and $\boldsymbol{u}^{\mathrm{FE}}$ are respective discretized variables, and $\boldsymbol{N}^{\lambda}$, $\boldsymbol{N}^{\mathrm{MLS}}$, and $\boldsymbol{N}^{\mathrm{FE}}$ are respective interpolation matrices:

$$
\begin{aligned}
L^{\mathrm{cons}} &= \int_{\Omega_b} (\mathbf{\Lambda})^{\mathrm{T}} \left(\boldsymbol{U}^{\frac{\mathrm{AP}}{\mathrm{FE}}} - \boldsymbol{U}^{\mathrm{FE}} \right) d\Omega \\
L^{\mathrm{cons}} &= \int_{\Omega_b} \left(\boldsymbol{N}^{\lambda}\boldsymbol{\lambda}\right)^{\mathrm{T}} \left(\boldsymbol{N}^{\mathrm{MLS}} \boldsymbol{u}^{\frac{\mathrm{AP}}{\mathrm{FE}}} - \boldsymbol{N}^{\mathrm{FE}} \boldsymbol{u}^{\mathrm{FE}} \right) d\Omega_b \\
L^{\mathrm{cons}} &= \boldsymbol{\lambda}^{\mathrm{T}} \left(\int_{\Omega_b} \boldsymbol{N}^{\lambda^{\mathrm{T}}} \boldsymbol{N}^{\mathrm{MLS}} d\Omega_b \boldsymbol{u}^{\frac{\mathrm{AP}}{\mathrm{FE}}} - \int_{\Omega_b} \boldsymbol{N}^{\lambda^{\mathrm{T}}} \boldsymbol{N}^{\mathrm{FE}} d\Omega_b \boldsymbol{u}^{\mathrm{FE}} \right) \\
L^{\mathrm{cons}} &= \boldsymbol{\lambda}^{\mathrm{T}} \left(\boldsymbol{G}^{\lambda/\mathrm{MLS}} \boldsymbol{u}^{\mathrm{AP/FE}} - \boldsymbol{G}^{\lambda/\mathrm{FE}} \boldsymbol{u}^{\mathrm{FE}} \right)
\end{aligned}
\tag{6.4}
$$

The Lagrange multiplier field is discretized using a similar mesh as the FEM domain. After adding the above constraint term in the overall system Lagrangian (Equation (6.2)), first variation of the Lagrangian can be taken with respect to the variables $\boldsymbol{\lambda}$, $\boldsymbol{u}^{\text{AP/FE}}$, and $\boldsymbol{u}^{\text{FE}}$ respectively to arrive at the following system of equations:

$$\begin{bmatrix} \boldsymbol{K}^{\text{FE}} & 0 & \boldsymbol{G}^{\lambda/\text{FE}^{\text{T}}} \\ 0 & \boldsymbol{K}^{S} & -\boldsymbol{G}^{\lambda/\text{MLS}^{\text{T}}} \\ \boldsymbol{G}^{\lambda/\text{FE}} & -\boldsymbol{G}^{\lambda/\text{MLS}} & 0 \end{bmatrix} \begin{Bmatrix} \boldsymbol{u}^{\text{FE}} \\ \boldsymbol{u}^{\text{AP/FE}} \\ \boldsymbol{\lambda} \end{Bmatrix} = \begin{Bmatrix} \boldsymbol{f}^{\text{FE,ext}} \\ \boldsymbol{T}^{\text{AP/MD}}\boldsymbol{u}^{\text{MD}} \\ 0 \end{Bmatrix} \tag{6.5}$$

The combined FEM and FEM anchor point system can be solved simultaneously at any given instance based on input MD time-averaged displacements $\boldsymbol{u}^{\text{MD}}$. The detailed derivation and explanation of the terms in Equations (6.4) and (6.5) can be found in references [5, 6, 10].

A second approach to constraining the FEM anchor points to the FEM domain is by directly "welding" the anchor point to the interior (e.g., the centroid) of the FEM 3-D element in the bridging domain. A detailed discussion on applying point forces to the interiors of FEM elements was provided in Chapter 2, which was based on the Dirac-delta function. That concept is key to applying spring forces via FEM anchor points to the interiors of FEM elements. The FEM anchor point displacements can be expressed in terms of the FEM nodal displacements as follows:

$$\boldsymbol{u}_i^{\text{AP/FE}} = \boldsymbol{N}^{\text{FE}}_{\boldsymbol{x}=\boldsymbol{x}_{0,i}^{\text{AP/FE}}}\boldsymbol{u}^{\text{FE}} \tag{6.6}$$

In the above, $\boldsymbol{N}^{\text{FE}}_{\boldsymbol{x}=\boldsymbol{x}_{0,i}^{\text{AP/FE}}}$ refers to the FEM interpolation function matrix evaluated at the coordinates of the ith FEM anchor point. Choosing an energy-based formulation and replacing the anchor point displacements from Equation (6.6) into Equation (6.3), the Lagrangian of the system can be expressed solely as a function of the FEM nodal displacements for a given input of MD anchor point displacements. By taking the variation of the Lagrangian with respect to the FEM nodal displacements, the following equation is obtained:

$$\left(\boldsymbol{K}^{\text{FE}} + \boldsymbol{N}^{\text{AP/FE}^{\text{T}}}\boldsymbol{K}^{S}\boldsymbol{N}^{\text{AP/FE}}\right)\boldsymbol{u}^{\text{FE}} = \boldsymbol{f}^{\text{FE,ext}} + \boldsymbol{N}^{\text{AP/FE}^{\text{T}}}\boldsymbol{K}^{S}\,\boldsymbol{T}^{\text{AP/MD}}\boldsymbol{u}^{\text{MD}} \tag{6.7}$$

The above equation can be solved to get the FEM nodal displacement vector $\boldsymbol{u}^{\text{FE}}$, for a given input of MD anchor point displacement. This, in turn, yields the FEM anchor point displacements, using Equation (6.6).

6.2.3 Overview of Iterative Algorithm for MD-FEM Coupling

In the previous subsections, individual aspects of the coupling algorithm, including bridging domain energy blending requirements and how the exchange of information takes places between FEM and MD domains via anchor points were described. In this subsection, the overall iterative algorithm that needs to be followed is described. A typical coupled simulation setup, with the FEM domain encompassing the MD domain, is shown in Figure 6.4.

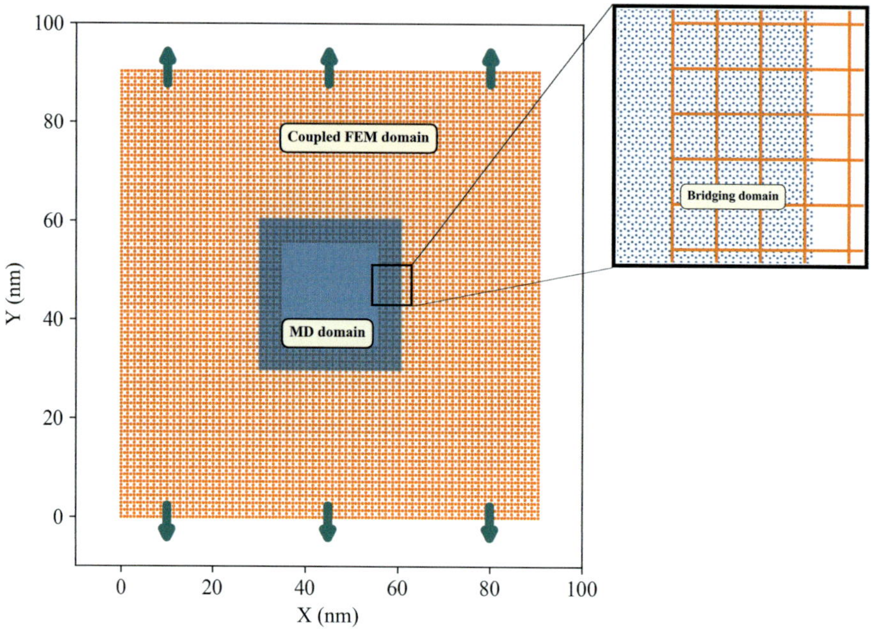

FIGURE 6.4 Typical MD-FEM coupling simulation domain, with bridging domain shown. Representative simulation case for uniaxial tension on monolayer graphene at $T = 300$ K.

The example geometry shown is for a uniaxial tension simulation on a monolayer graphene sheet at $T = 300$ K. Typically, far-field loading is applied to the FEM domain, and the local effects of that loading (potentially, near a crack tip) are studied in the coupled MD domain. Since an important objective of the present coupling method was to be able to use specialized MD solvers such as LAMMPS [9], without the need for significant modification, a "staggered" solution approach had to be followed, since the FEM and MD solvers are vastly different. In the staggered algorithm, when the FEM solver is executed, the atoms in the MD domain are kept frozen, and the last known positions of the MD anchor points are used as input to the FEM solver. On the other hand, when solution for the MD domain is carried out, the last known FEM anchor point positions are used as inputs, while the FEM domain is kept frozen. The MD solver runs for a specified amount of simulation time to allow equilibration of the MD system under the incremental loading. The updated time-averaged MD anchor positions are calculated and fed back to the FEM solver. This back-and-forth iteration between FEM and MD solvers can be described as a fixed-point iteration problem and continues until overall convergence in terms of forces and displacements is achieved in the bridging domain. The detailed iterative coupling algorithm is shown in the flowchart in Figure 6.5.

1. Initialization
   a. Define non-periodic MD domain.
   b. Define bridging domain and FEM mesh.
   c. Equilibrate MD domain with relevant thermostatting.
   d. Calculate initial time-averaged MD anchor point positions $\langle \boldsymbol{x}^{AP/MD} \rangle_\tau^{(0)}$.
   e. Precompute stiffness and coupling matrices.
2. Prescribe Dirichlet boundary conditions for current load step
   a. FEM Iteration *(i)*
      i. Input time-averaged MD anchor point displacements from previous iteration $\langle \boldsymbol{u}^{AP/MD} \rangle_\tau^{(i-1)}$.
      ii. Apply Dirichlet boundary conditions for the given loading step to the FEM domain and solve for $\boldsymbol{u}^{FE^{(i)}}$ and $\boldsymbol{u}^{AP/FE^{(i)}}$.
   b. MD Iteration *(i)*
      i. Input FEM anchor point displacements $\boldsymbol{u}^{AP/FE^{(i)}}$.
      ii. Run MD for required number of timesteps and compute $\langle \boldsymbol{u}^{AP/MD} \rangle_\tau^{(i)}$.
   c. Repeat until convergence criteria satisfied.
3. Increment loading and repeat steps 2 and 3 until maximum load level is reached.

FIGURE 6.5 Flowchart of the iterative "staggered" coupling algorithm for MD-FEM coupling.

6.3 EXAMPLES OF APPLICATION OF CONCURRENT MD-FEM COUPLING

Using the concurrent coupling approach that was described in the previous section, the results from example simulations on monolayer graphene and epoxy polymer are presented in this section.

6.3.1 Coupled MD-FEM Uniaxial Tension Simulations on Monolayer Graphene at T=300 K

In the first example simulation considered in this chapter, uniaxial tension simulations were performed on monolayer single-crystal graphene sheet at room temperature $T = 300$ K. The geometry and setup for the coupled MD-FEM simulation is shown in Figure 6.4. The sizes of the MD domain and the coupled FEM-MD domain are 30 × 30 nm and 90 × 90 nm respectively, with a 4.3 nm wide bridging domain. The sizes of the IVCs used were approximately 1.4 × 1.4 nm. All MD simulations were performed using LAMMPS [9], FEM simulations were performed using an adaptation of the NOVA-3D code developed by the authors [11], and a controller code written in Python was used for performing the iterative coupling. The atomistic force-field Adaptive Intermolecular Reactive Empirical Bond Order (AIREBO) [12] was used to model monolayer graphene. Details on other MD simulation-related parameters can be found in [10]. For the FEM domain, a linear elastic material model is used with Young's modulus and Poisson's ratio as input parameters, which are estimated from MD simulations and found to be 962 GPa and 0.25, respectively.

An essential aspect of the iterative coupling method is achieving convergence between MD and FEM at a given loading step. Two distinct convergence parameters were considered—the first related to reaction forces in the FEM domain and the second related to stresses in the coupled MD domain. The force-based convergence metric $e_i^{f,(k)}$ at a given iteration number k can be expressed as follows:

$$e_i^{f,(k)} = \frac{\|\boldsymbol{f}_i^{\text{cpl,ext},(k)} - \boldsymbol{f}_i^{\text{FE,ext}}\|_1}{\|\boldsymbol{f}_i^{\text{FE,ext}}\|_1}; \quad \Delta e_i^{f,(k)} = \frac{|e_i^{f,(k)} - e_i^{f,(k-1)}|}{e_i^{f,(k-1)}} \tag{6.8}$$

In the above equation, the reaction/external forces acting on the coupled FEM domain $\left(\boldsymbol{f}_i^{\text{cpl,ext},(k)}\right)$ are compared with reaction forces in a representative pure FEM simulation model $(\boldsymbol{f}_i^{\text{FE,ext}})$. Convergence is tested at an applied loading of $\varepsilon_{yy} = 2\%$ strain in the Y-direction as shown in Figure 6.4. The force convergence plots for the case of monolayer graphene using both the approaches for FEM anchor point definition are shown in Figure 6.6. Here, methods 1 and 2 refer to the "Direct coupling" approach defining FEM anchor and the "Lagrange multiplier" approach respectively, as described in the previous section. The final values of the error norms after 500 FEM-MD iterations are approximately 1%, and overall, both methods using FEM anchor point definition perform equally well, with direct approach proving to be more computationally efficient.

Another useful norm used to study convergence is a stress-based norm. In this, the stress in the coupled FEM and MD domains is compared with stresses in an equivalent pure MD system. FEM stresses are measured at the Gaussian integration points, whereas MD stresses are found from volume averaging of the atomistic virial stress at points across the MD domains. The error norm can be split into two parts as

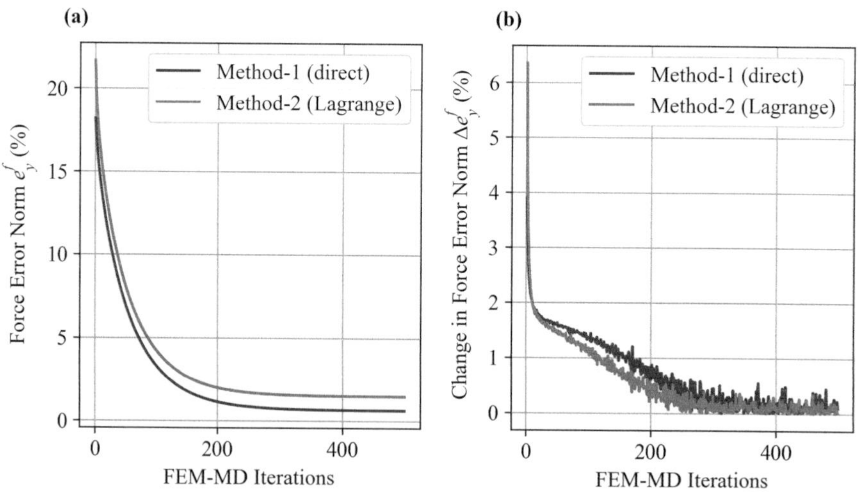

FIGURE 6.6 Force-based convergence norm plotted as a function of FEM-MD iteration number for monolayer graphene at $T = 300$ K. Method 1 corresponds to the "direct approach," and method 2 corresponds to the Lagrange multiplier method.

follows, with one for the coupled FEM domain $\left(e_{ij}^{\sigma,\text{FE},(k)}\right)$ and one for the coupled MD domain $\left(e_{ij}^{\sigma,\text{MD},(k)}\right)$ at a given iteration number k:

$$e_{ij}^{\sigma,\text{FE},(k)} = \frac{\left\|\boldsymbol{\sigma}_{ij}^{\text{cpl-FE},(k)} - \boldsymbol{\sigma}_{ij}^{\text{MD-FE}}\right\|_1}{\left\|\boldsymbol{\sigma}_{ij}^{\text{MD-FE}}\right\|_1} \tag{6.9a}$$

$$e_{ij}^{\sigma,\text{MD},(k)} = \frac{\left\|\boldsymbol{\sigma}_{ij}^{\text{cpl-MD},(k)} - \boldsymbol{\sigma}_{ij}^{\text{MD-MD}}\right\|_1}{\left\|\boldsymbol{\sigma}_{ij}^{\text{MD-MD}}\right\|_1} \tag{6.9b}$$

In the above, $\boldsymbol{\sigma}_{ij}^{\text{cpl-FE},(k)}$ and $\boldsymbol{\sigma}_{ij}^{\text{cpl-MD},(k)}$ are stresses in the coupled FEM and coupled MD domains, while $\boldsymbol{\sigma}_{ij}^{\text{MD-FE}}$ and $\boldsymbol{\sigma}_{ij}^{\text{MD-MD}}$ are equivalent stresses at the corresponding locations in the pure MD domain for comparison. The convergence behavior of the stress-based norms using both FEM anchor point definition methods is plotted in Figure 6.7. After 500 FEM-MD iterations, the errors in the coupled FEM stresses were 1%–2% and the errors in the coupled MD stresses were 2%–4%.

Based on the force- and stress-based convergence norms and the convergence plots in Figures 6.6 and 6.7, 500 FEM-MD iterations can be deemed to suffice for the given loading step in monolayer graphene, though the exact number of iterations required for an unknown problem will depend on various simulation parameters, including the coupling spring stiffness, with stiffer springs requiring more iterations for convergence. Once convergence is achieved for a given loading step, the stress distribution across the coupled MD-FEM domain may be evaluated and compared with results from a pure MD simulation. A plot of the stress σ_{yy} evaluated at an applied far-field load of $\varepsilon_{yy} = 2\%$ after convergence is achieved is shown in Figure 6.8.

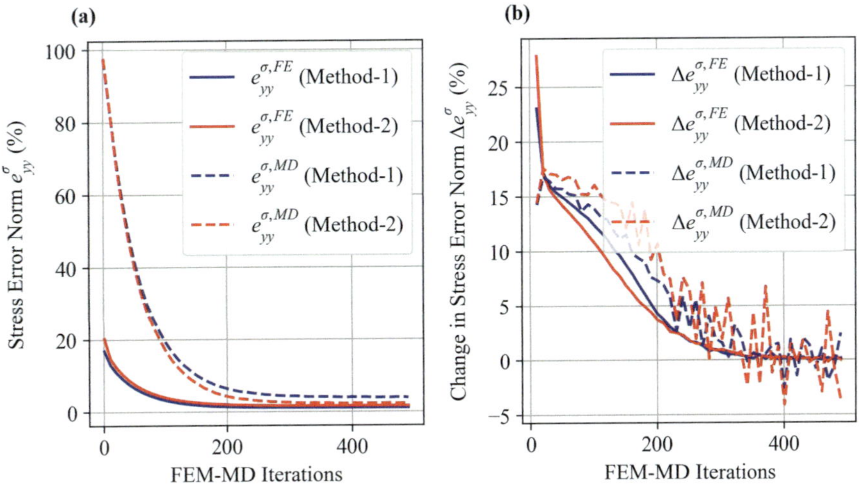

FIGURE 6.7 Stress-based convergence norms plotted as a function of FEM-MD iteration number for monolayer graphene at $T = 300$ K.

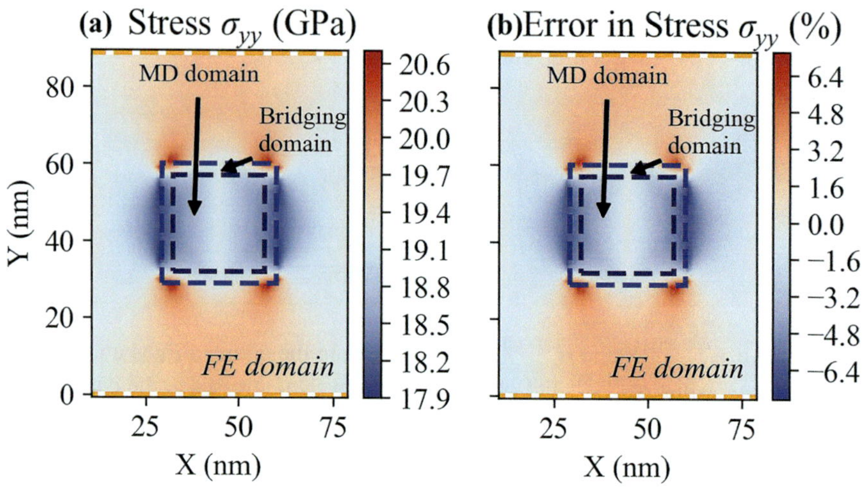

FIGURE 6.8 σ_{yy} stress contour plot (on left) for coupled MD-FEM domain in monolayer graphene at $T = 300$ K under an applied far-field load of $\varepsilon_{yy} = 2\%$. The percentage error in stress data is shown in the plot on the right. Spurious stress concentrations appear at the corners of the coupled MD domain.

In a perfect coupling scenario, there should be seamless transfer of stress, and for the uniaxial tension test considered here, stress should be uniform throughout the coupled MD-FEM domain. Practically, however, the stress distribution is nonuniform due to artifacts from the coupling method depending on the number of anchor springs used. Overall, the error is less than 8%, with the maximum error being at the bridging domain between the MD and FEM domains. Notably, there are stress concentrations at the corners of the rectangular coupled MD domain boundaries, and a reduction in stress near the left and right edges of the MD boundaries. Careful parametric explorations involving the coupling spring stiffness, bridging domain width, IVC size and number, and energy blending function can help to further reduce error in stress in the coupled MD-FEM domain.

6.3.2 Coupled MD-FEM Uniaxial Tension Simulations on Epoxy Polymer at T=1 K

In this example, the MD-FEM coupling method is applied to an amorphous material that is relevant to fiber-reinforced polymer composites—crosslinked epoxy resin. Specifically, the aerospace-grade epoxy resin EPON862/DETDA is simulated. The MD model for the crosslinked epoxy is based on the OPLS force field [13–15], which is a fixed bond force field. Crosslinking reactions, similar to the chemical reactions that take place in experiments, are simulated using the "fix bond/react" framework in LAMMPS, based on established methods in the literature [16, 17]. A 2-D planar geometry, similar to that used in the previous example for monolayer graphene, is used for epoxy and is shown in Figure 6.9.

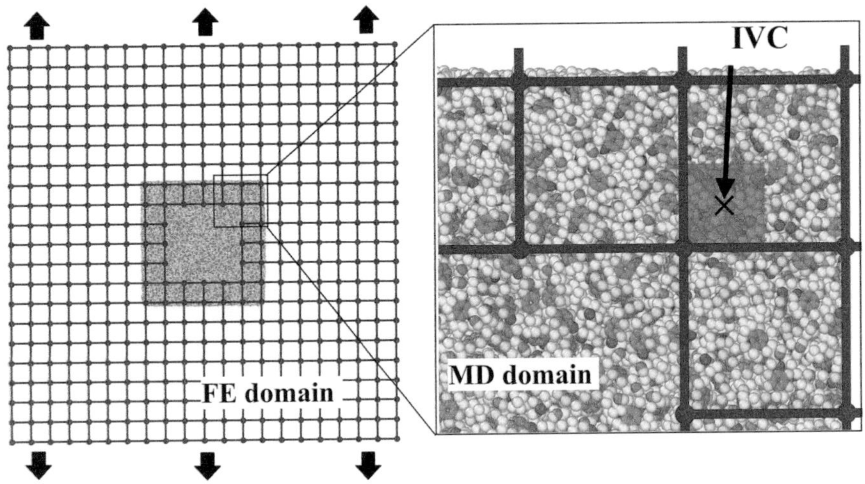

FIGURE 6.9 Schematic of coupled MD-FEM domain for performing uniaxial tension simulations on crosslinked epoxy polymer.

The MD domain and the coupled FEM-MD domain had sizes 20 × 20 nm and 60 × 60 nm respectively, with out-of-plane thickness 4 nm. For the epoxy polymer, simulations are performed at a temperature of $T = 1$ K, so that thermal fluctuations may be minimized while evaluating the results in the interest of clarity. A convergence analysis similar to the monolayer graphene case was performed based on the force and stress norms at an applied loading of $\varepsilon_{yy} = 1\%$. At this loading level, the Young's modulus of the epoxy resin, based on additional MD characterization tests, was found to be 8 GPa. It should be noted that epoxy resin exhibits nonlinear behavior and therefore the secant modulus varies with strain. The convergence results from the FEM and MD stress norms in Equations (6.9a) and (6.9b) respectively are plotted against MD-FEM iteration number in Figure 6.10.

Based on the convergence plots, 100 MD-FEM iterations were found to be sufficient in ensuring converged stresses for epoxy polymer at $T = 1$ K. At the converged state, the σ_{yy} stresses in the coupled MD-FEM domain were evaluated at the applied loading level of $\varepsilon_{yy} = 1\%$ and plotted in Figure 6.11.

In contrast to the monolayer graphene case, there is significant spatial nonuniformity in the material properties and σ_{yy} stress distribution in crosslinked polymer, especially at the nanoscale considered here. The "error in stress" contour plot in Figure 6.11 compares the stresses in the coupled MD domain with those in an equivalent pure MD domain. The error in stress is less than 11%, with stress concentration again at the corners of the MD domain. The central region of the coupled domain exhibits reduced stress, which can likely be alleviated by using stiffer coupling springs, albeit at the cost of increased MD-FEM iterations required for convergence. Beyond evaluating stresses at a single loading point, the MD-FEM coupling method can be used to explore interesting nonlinear constitutive behavior that the epoxy polymer exhibits at higher strains. The nonlinear material behavior up

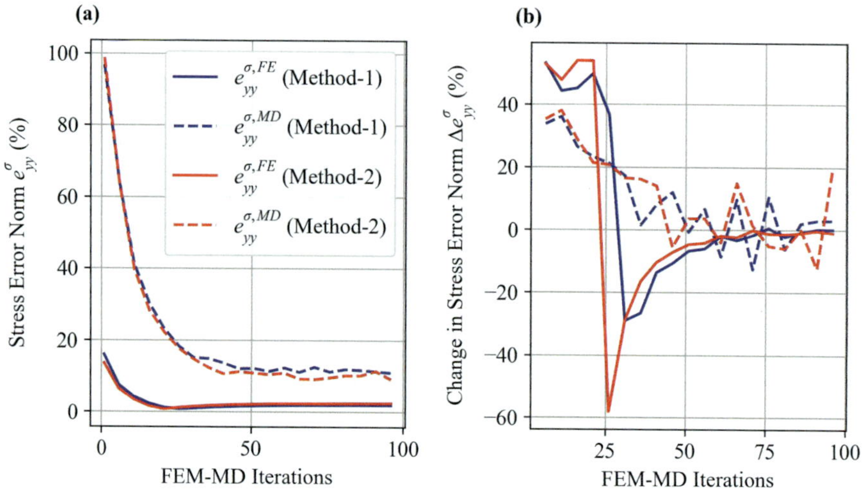

FIGURE 6.10 Stress-based convergence norms plotted as a function of FEM-MD iteration number for epoxy polymer at $T = 1$ K.

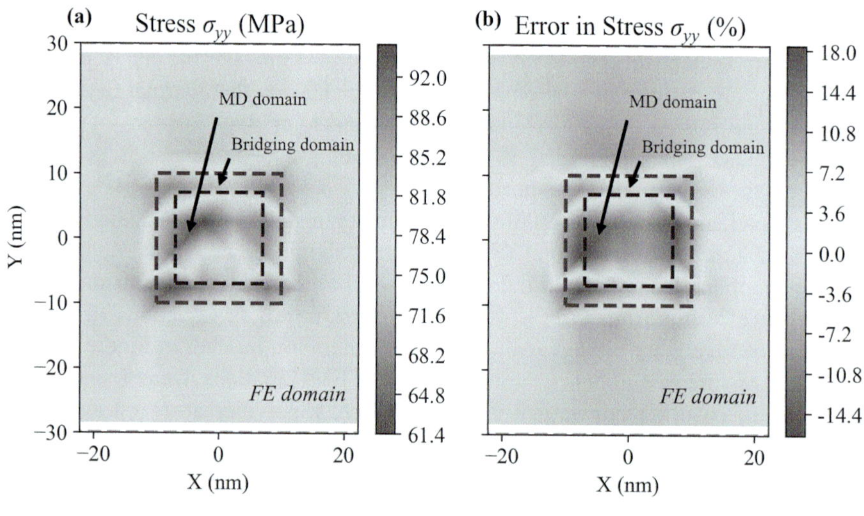

FIGURE 6.11 σ_{yy} stress contour plot (left) for coupled MD-FEM domain in epoxy polymer at $T = 1$ K under an applied far-field load of $\varepsilon_{yy} = 1\%$. The percentage error in stress contour is shown on the right.

to 9% strain is fitted to a nonlinear Ramberg-Osgood-type material model, which may be expressed as follows in terms of stress σ, strain ϵ, and material parameters E, K, and n:

$$\epsilon = \frac{\sigma}{E} + K\left(\frac{\sigma}{E}\right)^n \tag{6.10}$$

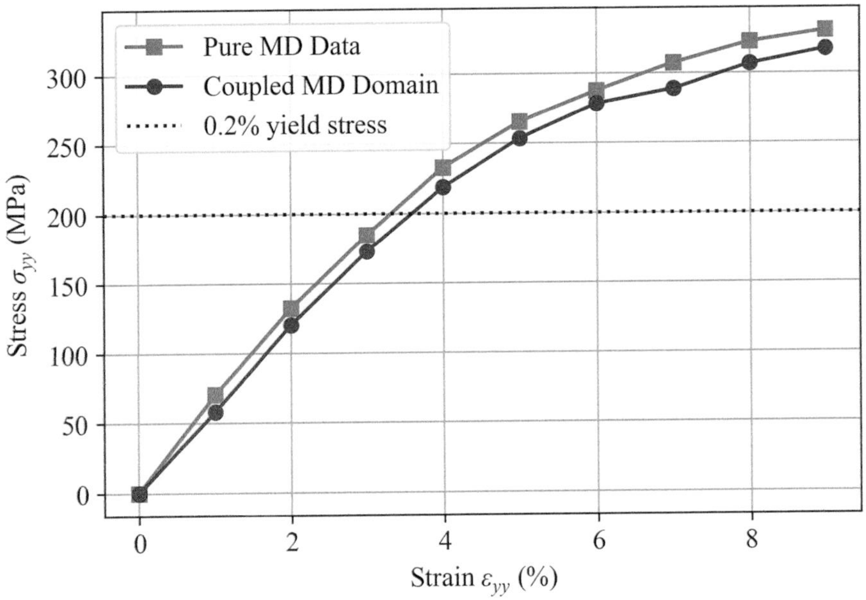

FIGURE 6.12 Average stress in coupled MD domain for coupled uniaxial tension test on epoxy polymer based on Ramberg-Osgood material model in coupled FEM domain at $T = 1$ K, compared with pure MD domain for reference.

The Ramberg-Osgood material model is then used in the coupled FEM solver and coupled iterations between the FEM and MD domains are performed with incremental loading steps of $\Delta\varepsilon_{yy} = 1\%$. The average stress in the coupled MD domain is then plotted in Figure 6.12 as a function of applied far-field strain and compared with the average stress in the pure MD domain (used for reference).

The plot shows that there is excellent agreement between the coupled MD-FEM and pure MD results with errors less than 6% even at high levels of applied strain, thereby validating the overall coupling methodology, even though local stress concentrations are not accurately simulated in this simulation.

6.4 SUMMARY

In this chapter, concurrent coupling between MD and FEM simulations was described in connection with the theory of transition flaw size and the K-test approach. Next, a modern framework for performing practical concurrent coupled simulations, agnostic of material type or MD force field, that can make use of specialized MD solvers like LAMMPS, was introduced. Essential theoretical foundations of the coupling method and practical nuances when attempting to implement the method were discussed. Two methods were described: the first one is based on Lagrange multipliers as used in the Arlequin method and the Capriccio method, and the second one is a more direct point-based approach developed by the authors.

Finally, the concurrent coupling method was applied to study example problems for monolayer graphene (crystalline material) and epoxy polymer (amorphous material). Uniaxial tension tests were performed on the example systems using the concurrent coupling method, iterative convergence was tested, and finally the resultant stress distributions and errors were assessed. While global stress vs. strain was accurately computed by the MD-FEM coupled system, nonuniformity in the local stress field was observed near the coupling domain due to the lack of a sufficient number of IVCs. This chapter paves the way for more advanced applications of concurrent coupling related to simulation of fracture problems in subsequent chapters.

REFERENCES

1. Shenoy, V., et al., Quasicontinuum models of interfacial structure and deformation. *Physical Review Letters*, 1998. **80**(4): p. 742.
2. Xiao, S. and T. Belytschko, A bridging domain method for coupling continua with molecular dynamics. *Computer Methods in Applied Mechanics and Engineering*, 2004. **193**(17–20): p. 1645–1669.
3. Shilkrot, L., R.E. Miller, and W. Curtin, Coupled atomistic and discrete dislocation plasticity. *Physical Review Letters*, 2002. **89**(2): p. 025501.
4. Miller, R.E. and E.B. Tadmor, A unified framework and performance benchmark of fourteen multiscale atomistic/continuum coupling methods. *Modelling and Simulation in Materials Science and Engineering*, 2009. **17**(5): p. 053001.
5. Pfaller, S., A. Kergaßner, and P. Steinmann, Optimisation of the capriccio method to couple particle-and continuum-based simulations of polymers. *Multiscale Science and Engineering*, 2019. **1**: p. 318–333.
6. Pfaller, S., et al., An Arlequin-based method to couple molecular dynamics and finite element simulations of amorphous polymers and nanocomposites. *Computer Methods in Applied Mechanics and Engineering*, 2013. **260**: p. 109–129.
7. Saether, E., V. Yamakov, and E.H. Glaessgen, An embedded statistical method for coupling molecular dynamics and finite element analyses. *International Journal for Numerical Methods in Engineering*, 2009. **78**(11): p. 1292–1319.
8. Bauman, P.T., et al., On the application of the Arlequin method to the coupling of particle and continuum models. *Computational Mechanics*, 2008. **42**: p. 511–530.
9. Thompson, A.P., et al., LAMMPS-a flexible simulation tool for particle-based materials modeling at the atomic, meso, and continuum scales. *Computer Physics Communications*, 2022. **271**: p. 108171.
10. Aditya, S.S., T. Sohail, and S. Roy, An atomistic-continuum concurrent statistical coupling technique for amorphous materials using anchor points. *Modelling and Simulation in Materials Science and Engineering*, 2023. **31**(7): p. 075012.
11. Roy, S. and J. Reddy, Finite-element models of viscoelasticity and diffusion in adhesively bonded joints. *International Journal for Numerical Methods in Engineering*, 1988. **26**(11): p. 2531–2546.
12. Stuart, S.J., A.B. Tutein, and J.A. Harrison, A reactive potential for hydrocarbons with intermolecular interactions. *The Journal of Chemical Physics*, 2000. **112**(14): p. 6472–6486.
13. Dodda, L.S., et al., LigParGen web server: an automatic OPLS-AA parameter generator for organic ligands. *Nucleic Acids Research*, 2017. **45**(W1): p. W331–W336.
14. Dodda, L.S., et al., 1.14* CM1A-LBCC: localized bond-charge corrected CM1A charges for condensed-phase simulations. *The Journal of Physical Chemistry B*, 2017. **121**(15): p. 3864–3870.

15. Jorgensen, W.L. and J. Tirado-Rives, Potential energy functions for atomic-level simulations of water and organic and biomolecular systems. *Proceedings of the National Academy of Sciences of the United States of America*, 2005. **102**(19): p. 6665–6670.
16. Odegard, G.M., et al., Predicting mechanical response of crosslinked epoxy using ReaxFF. *Chemical Physics Letters*, 2014. **591**: p. 175–178.
17. Li, C. and A. Strachan, Molecular dynamics predictions of thermal and mechanical properties of thermoset polymer EPON862/DETDA. *Polymer*, 2011. **52**(13): p. 2920–2928.

7 Modeling of Fracture in Crystalline Materials Using MD and Extraction of Cohesive Zone Parameters

7.1 INTRODUCTION

An important crystalline material that has spurred significant research interest in recent years is monolayer graphene, a 2-D allotrope of carbon with sp^2 bonding. The experimentally measured fracture toughness of monolayer graphene is 15.9 J/m^2 [1]. Section 7.2 describes the application of the atomistic J-integral to obtain the cohesive traction-separation law for crystalline monolayer graphene sheets with a center crack that is crack-size independent. The primary objective here is to demonstrate the feasibility as well as the correct methodology using MD for obtaining cohesive traction-separation laws at the nanoscale that is a true material property. Section 7.3 describes the application of FEM-MD concurrent coupling to predict the fracture toughness of monolayer graphene.

7.2 FRACTURE SIMULATION IN GRAPHENE: EXTRACTION OF COHESIVE TRACTION-SEPARATION LAW

7.2.1 Center-Cracked Fracture Simulation on Monolayer Graphene (Uncoupled Case)

It is now well-established that the cohesive traction-separation law for any given material should be crack-length independent, as it is a material property. In this context, researchers have obtained the traction-separation law from the far-field stress-strain response of a sheet of a material (or a bi-material interface) using MD, in some cases with a flaw or a crack in it [2–4], and in other cases without a flaw [5, 6]. On the other hand, some researchers have used stress-displacement data from the near-tip response of a sheet of a material (or a bi-material interface) to obtain the traction-separation law when a crack is present [2, 7, 8]. In this context, Dewapriya et al. [4] discussed that the critical far-field stress in the fracture of a single graphene sheet is strongly crack-length dependent for both the zig-zag and armchair configurations. By obtaining the traction-separation law at five different locations relative to the crack plane, Zhou et al. [2] studied the effect of stress and displacement measurement

DOI: 10.1201/9781003440734-7

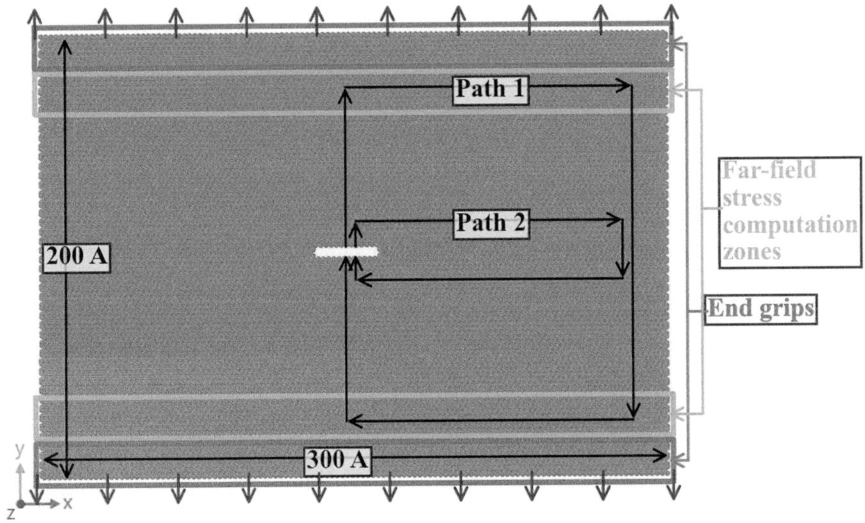

FIGURE 7.1 *J*-integral contours showing far-field (Path 1) and near-tip (Path 2) contour paths in monolayer graphene.

location relative to the crack plane for obtaining the traction-separation law for Mode-I as well as mixed-mode crack at the interface between two brittle materials using MD simulations. By means of an extensive sensitivity study, they were able to demonstrate that the cohesive traction-separation law becomes increasingly crack-length independent only when the measurements are taken very close to the crack plane, of the order of 10–20 Å above and below the crack plane. However, none of the above papers discuss the exact mathematical reasoning as to why such a large discrepancy exists between far-field and near-tip traction-separation data. The objective of this section is to investigate the effect of far-field data compared with near-tip data and establish a mechanics-based mathematical explanation for this discrepancy that is valid even at the nanoscale.

A large number of researchers have obtained the cohesive traction-separation law for a sheet of 2-D material from the far-field stress-strain response. Figure 7.1 shows two concentric paths for computing the *J*-integral for a center-cracked graphene sheet. Of course, as discussed in Chapter 4, the *J*-integral remains path independent by definition even at the nanoscale and must have the same value on all concentric paths [9, 10], in the absence of inelastic dissipation. When cohesive traction-separation law is obtained from far-field data in an MD simulation, typically the virial stresses in the boxed regions (in red), as depicted in Figure 7.1, are volume averaged at every time step and plotted as a function of the separation distance between the top and bottom boxes of the full simulated test specimen.

7.2.1.1 *J*-integral Computation for Path 1

If we assume that P_{ij} are Cartesian components of the First Piola-Kirchhoff stress tensor in the red-boxed region, and H_{ik} are the corresponding components of the displacement gradient tensor in the far-field, then invoking Equation (4.39) over contour Path 1,

$$\int_{\Gamma_1} \langle -P_{ij}H_{ik}\rangle N_j d\Gamma = (J)_k - \int_{\Gamma_1} \langle \Psi\delta_{kj}\rangle N_j d\Gamma \tag{7.1}$$

where, i, j, k ranges between 1 and 2, and Γ_1 implies contour Path 1 shown in Figure 7.1, and $\langle * \rangle$ implies ensemble average. Mathematically, this implies that for contour Path 1 we can set $k = 1$ to obtain the component of J in the X_1 direction, i.e., $(J)_1$ in Equation (7.1). Further, if we divide the contour Path 1 into vertical and horizontal segments, we recognize that on the vertical segments the unit normal $N_1 = 1$ and $N_2 = 0$. Conversely, along the horizontal segments of the contour Path 1 shown in Figure 7.1, the unit normal $N_1 = 0$ and $N_2 = 1$. Further, if it is assumed that the deformations are relatively small within the contour, we can set $P_{22} = \sigma_{22}$, which is the corresponding Cauchy stress component in the loading direction, and set $i = 2$, $j = 2$ and $k = 1$ in Equation (7.1), giving the far-field traction-separation law,

$$\begin{aligned}\int_{\Gamma_{1\text{-horizontal}}} \left\langle -\sigma_{22}\frac{\partial u_2}{\partial x_1}\right\rangle dx_1 &= \int_0^{u_{2\,\max}} \langle -\sigma_{22}\rangle du_2^{\text{top}} + \int_{u_{2\,\max}}^{0} \langle -\sigma_{22}\rangle du_2^{\text{bottom}} \\ &= \int_0^{u_{2\,\max}} \langle \sigma_{22}\rangle d\Delta u_2 = (J)_1 - \int_{\Gamma_{1\text{-vertical}}} \langle \Psi\rangle d\Gamma \end{aligned} \tag{7.2}$$

where, $\Delta u_2 = u_2^{\text{botom}} - u_2^{\text{top}}$ is the relative separation between the bottom and top horizontal portions of the contour Path 1 measured along the horizontal length of the contour in the negative X-direction from the free edge of the specimen toward the crack tip. From Equation (7.2) it becomes immediately apparent that, at crack initiation, the traction-separation law described on the right-hand side (RHS) of this equation is not equal to the fracture energy (J_c or G_c) due to the presence of the Helmholtz free energy term (Ψ) on the RHS. The traction-separation law obtained in this manner shows crack size dependence because the free-energy term changes with the increase in crack length, and hence, is no longer a material property. This issue is discussed further in the next section based on MD simulations for a single-crystal graphene sheet with a center crack.

7.2.1.2 *J*-integral Computation for Path 2

However, if the contour path height (h) is made sufficiently small as schematically shown in Figure 7.2, the Helmholtz free energy term can be reduced to a negligibly small value in the J-integral equation for Mode-I as well as for mixed mode loadings.

Mathematically, this implies that for the narrow contour we can set $k = 1$ to obtain the component of J in the X_1 direction, i.e., $(J)_1$ in Equation (7.1), and $j = 2$ to define the unit normal N_2 along the narrow cohesive-contour shown in the schematic in Figure 7.2 (and also Path 2 in Figure 7.1). Note that $N_1 = 0$ for this entire horizontal contour path, with $k = 1, j = 2$, and as a result $\delta_{kj} = \delta_{12}$.

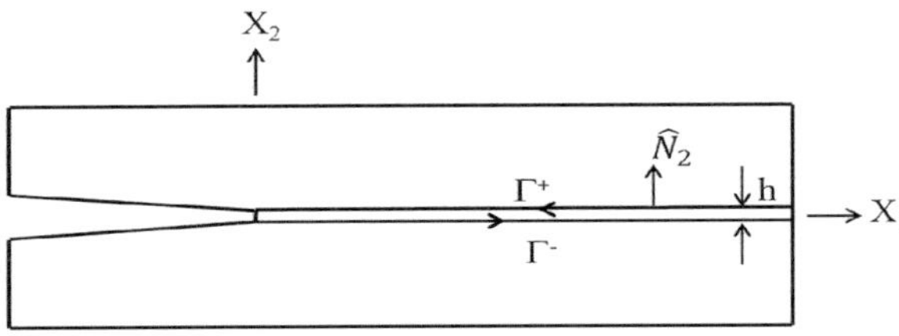

FIGURE 7.2 Schematic of J-integral contour for obtaining near-tip traction-separation law.

$$(J)_k = \int_{\Gamma} \langle \Psi \delta_{kj} - P_{ij} H_{ik} \rangle N_j d\Gamma \tag{7.3}$$

Then the free energy term on the RHS of Equation (7.3) vanishes because $\delta_{12} = 0$, from definition of Kronecker delta. Therefore, the only terms that need to be evaluated along the narrow cohesive contour are the Cartesian components of the first Piola-Kirchhoff stress tensor (P_{i2}), and components of the displacement gradient tensor (H_{i1}), giving

$$\int_{\Gamma_2} \langle -P_{i2} H_{i1} \rangle N_2 d\Gamma = \int_{\Gamma_2} \langle -P_{12} H_{11} - P_{22} H_{21} \rangle N_2 d\Gamma = (J)_1 \tag{7.4}$$

where, Γ_2 implies contour Path 2 shown in Figure 7.2. While Equation (7.4) is fully capable of including mixed-mode fracture with shear stress present as one might expect at a bi-material interface, the derivation is restricted to pure Mode-I loading of a single sheet of material, in which case the shear stress term, P_{12}, vanishes, giving,

$$\int_{\Gamma_2} \left\langle -P_{22} \frac{\partial u_2}{\partial x_1} \right\rangle N_2 d\Gamma = (J)_1 \tag{7.5}$$

Similar to the derivation for Path 1, assuming that the deformations are relatively small within the contour, we can set $P_{22} = \sigma_{22}$, the Cauchy stress, and unit normal $N_2 = 1$ over the contour, giving the near-tip traction-separation law,

$$\int_{0}^{u_{2\,\max}} \langle \sigma_{22} \rangle d\Delta u_2 = (J)_1 \tag{7.6}$$

where $\Delta u_2 = u_2^{\text{botom}} - u_2^{\text{top}}$ is again the separation distance between the bottom and the top segments of the contour measured along the length of the narrow contour. From Equation (7.6) it becomes immediately apparent that, at crack initiation, the traction-separation law described on the left-hand side of this equation is equal to the fracture energy (J_c or G_c) independent of crack length, because of the absence of the Helmholtz free energy term present in Equation (7.2). For contours for which h is large, such as traction-separation law formulated based on a contour taken far-field from the crack, there will be significant error induced due to the free energy term present as shown in Equation (7.2). In such cases, the area under the traction-separation law is no longer equal to the fracture energy because the Helmholtz free energy term cannot be ignored, and therefore the work of separation is no longer purely a material property, but crack-size dependent. Consequently, the traction-separation law obtained from far-field data does not have a constant area under the curve and exhibits crack-length dependence as discussed in the next section. The analytical derivation presented above is quite general and is applicable for crystalline materials, such as graphene, as well as for amorphous materials, such as epoxy polymers, and even for bi-material interfaces. However, care should be taken to ensure that the fully developed fracture process zone is contained inside the J-contour in order to maintain path-independence of J.

7.2.2 Model Verification Using Molecular Dynamics: Far-field vs. Near-tip Traction-Separation Law

For verification of the concepts proposed in Section 7.2.1, a 300 Å × 200 Å × 3.14 Å single-crystal graphene sheet with a center crack, as shown in Figure 7.1, was modeled using the Adaptive Intermolecular Reactive Empirical Bond Order (AIREBO) potential [10] available in the open-source LAMMPS MD software. Unlike in Section 4.8, where ReaxFF was used to model crack initiation and growth in a center-cracked monolayer graphene, the use of the AIREBO potential instead of ReaxFF allows the MD simulations to be less computationally intensive, and yet allow for bond breakage. The width of the sheet was intentionally chosen as 300 Å in order to exceed the critical strip width for flaw tolerance for a single-crystal graphene sheet (~170 Å) as defined in Zhang et al. [11], and also to be at least 10 times greater than the largest half-crack length to avoid edge effects. The simulations were performed at $T = 1$ K to reduce stress oscillations due to thermal motion of the carbon atoms; however, the results obtained should be valid at elevated temperatures as well. The model contained 24,400 atoms in the simulation volume. Periodic boundary conditions were applied along the $x-$ and y-directions. Energy minimization was performed using the conjugate gradient method followed by ensemble equilibration at 1 K and 0 bar pressure under the NPT ensemble, performed for 50 ps with 0.5 fs time steps, controlled by a Nose-Hoover barostat.

A cutoff radius of 2 Å was used for the AIREBO potential for graphene to ensure stress convergence. A uniform displacement was applied along the far-field edges of the graphene sheet in the armchair chiral direction as shown by the green boxes in Figure 7.1, and was increased in regular steps of 0.5 ps for 1000 steps, followed by a hold time of 500 ps to allow for energy equilibration between load increments. The average strain rate applied is 1/ns. In this context, previous work by the authors has shown that single-crystal graphene is relatively insensitive to strain rate as long as the relaxation ratio remains constant [9]. The displacement was ramped up in steps until crack initiation occurred. Crack initiation was determined by monitoring the bond-weight function between the pair of carbon atoms right at the crack tip.

The bond-weight function is used in the AIREBO potential to indicate when a bond has separated, i.e., the bond-weight is 1 when the bond is intact, and 0 when the bond is broken [12]. Therefore, crack initiation was assumed to occur when the bond weight was reduced to zero. In this manner, crack initiation for seven different half crack lengths was simulated, starting with the pristine case (no crack), followed by $a = 2.5, 4.9, 7.6, 10.3, 15.2, \text{and } 17.7$ Å, respectively. For each case, the average far-field stress was obtained by volume averaging the virial stress in the y-direction for atoms in the region within the red box in Figure 7.1, and plotted as a function of the average separation between the atoms in the red boxes at the top and bottom of the figure. In order to assure quasi-static crack initiation, simulations were stopped at the moment of crack initiation for each crack-length case to avoid dynamic crack growth effects and/or lattice trapping.

Figure 7.3 shows the far-field virial stress in the y-direction plotted as a function of the far-field separation for each half crack-length case. It is evident from this plot that the traction-separation law obtained using far-field data is clearly crack length dependent. Table 7.1 lists the critical far-field fracture stress as a function of crack length, as well as the work of separation obtained by integrating the area under the curve. It can be readily observed that there is a strong crack-length dependence to these data due to the free-energy term present in Equation (7.2). Hence, the

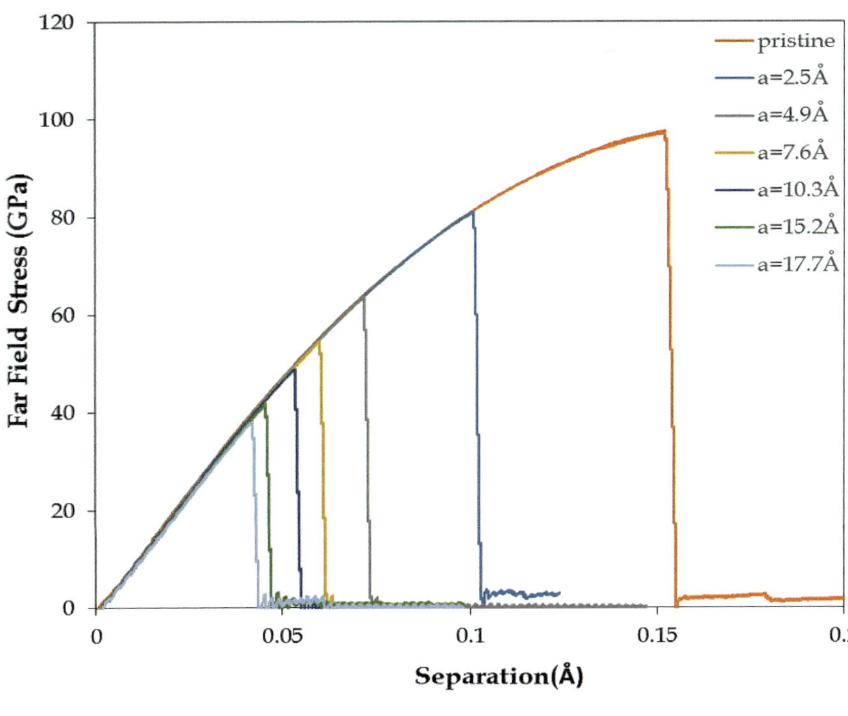

FIGURE 7.3 Far-field traction vs. separation for different half crack lengths in graphene using far-field data.

TABLE 7.1
Fracture Strength and Fracture Energy for Different Crack Lengths in Graphene Using Far-field Data

Half Crack Length (Å)	Fracture Strength (GPa)	Fracture Energy (J/m^2)
0	99.77	11.72
2.5	81.93	4.55
4.9	64.40	2.42
7.6	55.39	1.69
10.3	49.75	1.35
15.2	42.69	0.97
17.7	39.05	0.82

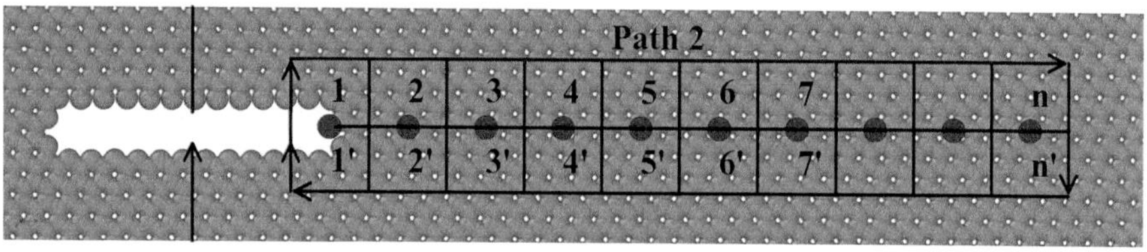

FIGURE 7.4 Computation boxes used for the computation of near-tip traction-separation law.

traction-separation law and fracture energy obtained in this manner are structural properties, and not material properties, as summarized in Table 7.1.

In order to accurately obtain the near-tip traction-separation law, the region ahead of the crack tip bounded by Path 2 in Figure 7.1 was subdivided into computation boxes as shown in Figure 7.4.

The virial stress at the moment of crack initiation computed within each box above and below the crack plane, for example, 1 and 1′, was averaged and plotted as a function of its x-location from the crack tip, as shown in Figure 7.5 for half-crack lengths of 2.5, 7.6, and 17.7 Å, respectively.

The critical far-field stress at fracture initiation is also plotted as a dashed red line for each case. As can be observed from Figure 7.5, notch sensitivity, defined as the ratio between the maximum crack-tip stress and the far-field stress, increases rapidly with crack size for graphene. The separation distance between each pair of atoms along the plane of the crack, shown in Figure 7.4 as blue dots for a generic pair of atoms, was also obtained as a function of the x-location from the crack tip. Finally, these two data sets were combined to eliminate the x-position coordinate and to construct the traction-separation law from the near-tip data as shown in Figures 7.6–7.8.

Figures 7.6–7.8 show near-tip traction-separation law for selected cases with half-crack lengths of 2.5, 7.6, and 17.7 Å, respectively. As can be observed from the identical shapes of the plots, any dependence on crack length is noticeably absent

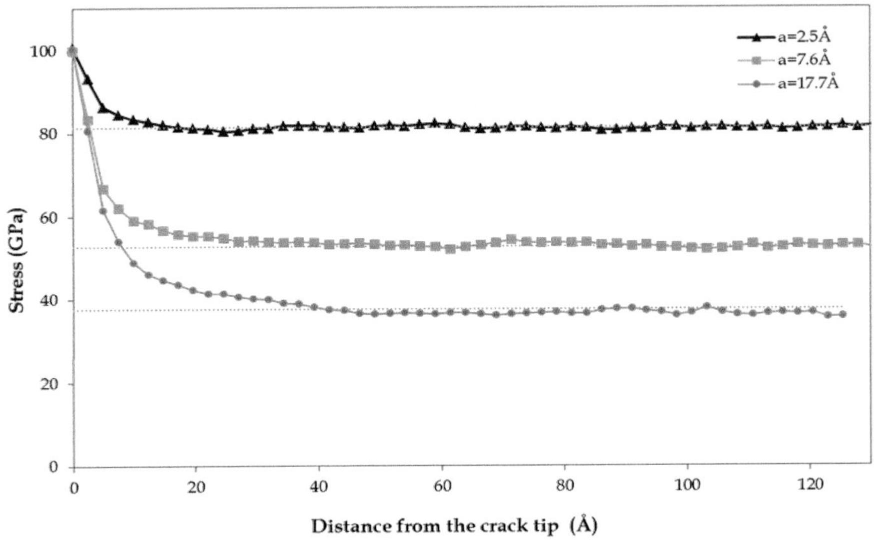

FIGURE 7.5 Near-tip traction plotted as a function of distance from crack tip for different half-crack lengths. The critical far-field stress at fracture is also plotted as red dashed line for each case to show increasing notch-sensitivity effect with increasing crack size.

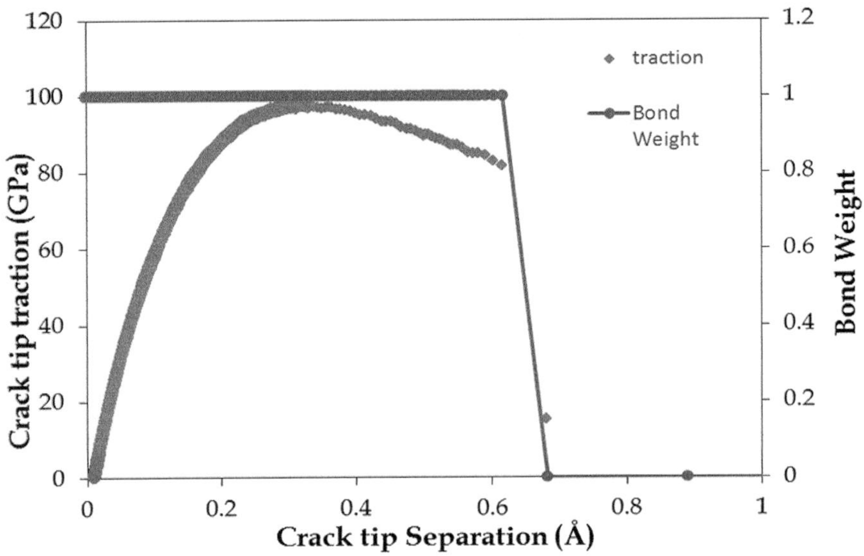

FIGURE 7.6 Near-tip traction-separation law for half crack length = 2.5 Å case. The bond-weight function is also plotted to determine the exact point of crack initiation when bond order goes to zero.

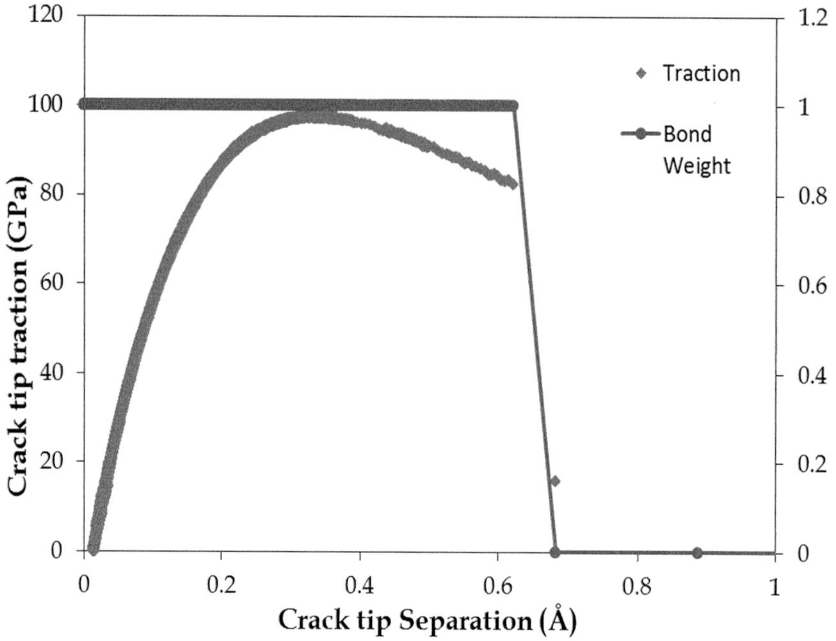

FIGURE 7.7 Near-tip traction-separation law for half crack length = 7.6 Å case. The bond-weight function is also plotted to determine the exact point of crack initiation when bond order goes to zero.

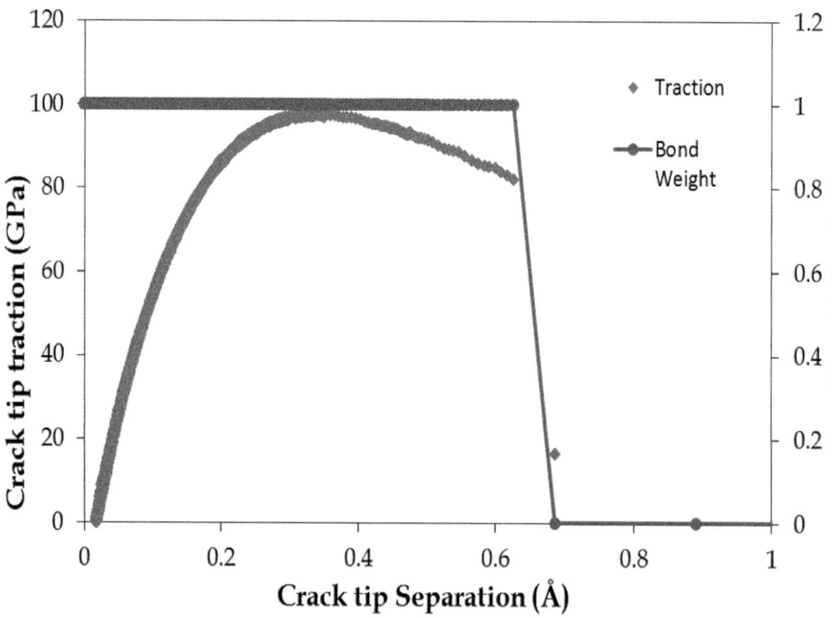

FIGURE 7.8 Near-tip traction-separation law for half crack length = 17.7 Å case. The bond-weight function is also plotted to determine the exact point of crack initiation when bond order goes to zero.

from the near-tip traction-separation plots, thereby implying that it is truly a material property. Further, for each traction-separation plot, the bond-weight function was also superposed to determine the exact point at which the crack tip atoms separate, thereby causing crack initiation.

The use of the bond-weight function provides a direct means of determining the exact point of bond separation, i.e., crack initiation, and is deemed more accurate than bond failure being assumed to occur at a 15% strain level as described in [4]. As can be seen from Figures 7.6–7.8, the bond-weight function plotted on the right side of the ordinate clearly indicates that the traction-separation law for graphene reaches a peak value and then undergoes a nonlinear softening phase before final debonding occurs. This is in direct contrast to the far-field traction-separation data that shows mostly linear/hardening behavior until crack initiation (see Figure 7.3).

Table 7.2 lists the peak cohesive stress as well as the fracture energy for the near-tip data. The constant fracture energy of 9.75 J/m^2 obtained from the area under the traction-separation curve agrees reasonably well with published data [13] for the failure of a graphene sheet being loaded in the armchair chiral direction at $T = 1$ K. Tensile loading in the zig-zag chiral direction produces similar results for near-tip data and hence is not presented here in the interest of space.

Maximum cohesive stress vs. crack length for the near-tip case for armchair chirality is summarized in Figure 7.9 and clearly shows the absence of any dependence on crack length, except for the pristine case when $a = 0$. From Figure 7.9 it is evident that the rupture strength of the graphene sheet is 100 GPa in the absence of a flaw or a crack, and it is slightly higher than the fracture strength of around 98 GPa when a crack is present. Consequently, using rupture data from uncracked specimens to obtain the cohesive traction-separation law can lead to nonconservative estimates of fracture strength. Also evident from Table 7.2 is the fact that the influence of the ratio of notch radius ($R = 0.267$ nm) to crack length on fracture energy is minimal for the cases investigated in this paper. However, Table 7.2 (Column 4) indicates that, based on our MD results, the cohesive zone length (i.e., FPZ) decreases with increasing crack length even though the cohesive traction-separation law remains unchanged. Because the cohesive zone length is measured from MD data as the

TABLE 7.2
Cohesive Fracture Strength, Fracture Energy, and Process Zone Size of a Center-Cracked Graphene Sheet for Different Crack L

Half Crack Length (Å)	Max Stress at the Crack Tip (GPa)	Fracture Energy (J/m^2)	Cohesive Zone Length (Å)
2.5	97.65	9.759	7.38
4.9	97.86	9.745	5.35
7.6	97.82	9.743	2.46
10.3	97.94	9.736	2.35
15.2	97.7	9.734	1.64

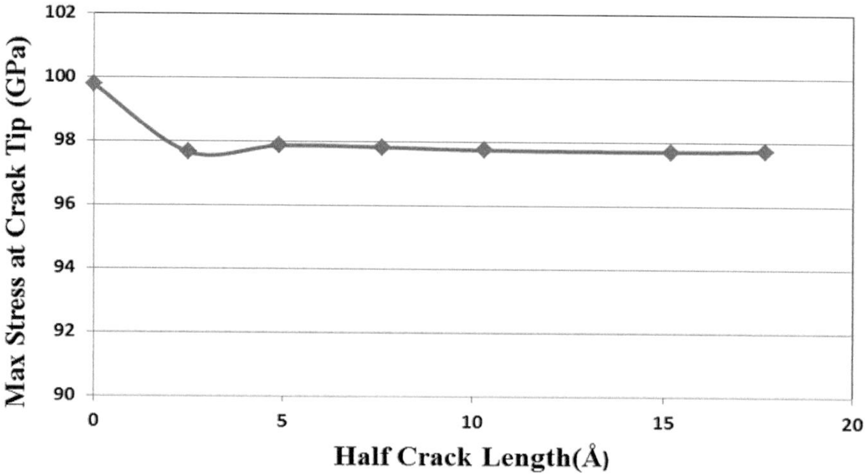

FIGURE 7.9 Variation of peak cohesive stress in graphene at the crack tip for different crack lengths, along with the pristine (no-crack) case.

distance between the location of the peak cohesive stress and the crack tip as shown in Figure 7.4, a decrease in cohesive zone length indicates a decrease in the size of FPZ, and therefore, a transition from quasi-brittle to fully brittle failure with increasing crack length in single-crystal graphene. This observation fully corroborates the critical crack length concept for crystalline materials.

7.2.3 Derivation of Traction-Separation Law from Potential Energy Consideration

Some researchers have employed the change in the overall potential energy of an MD system, before and after crack initiation, to obtain the traction-separation law [13, 14]. They assumed that in the absence of dissipative mechanisms, the total work done on a system per unit volume is $\Delta W(\varepsilon) = \frac{1}{2}\sigma\Delta\varepsilon$, where σ is the uniaxial stress applied on the system and $\Delta\varepsilon$ is the strain due to the applied stress. Then, defining total work done (ΔW) equal to the change in potential energy $(\Delta\Pi)$ obtained from MD as a function of separation distance, the volume averaged traction-separation law can be obtained as,

$$\sigma(\Delta\varepsilon) = \frac{2(\Pi_{\text{final}} - \Pi_{\text{initial}})}{V\Delta\varepsilon} \tag{7.7}$$

where V is the volume of the system (in this case, the graphene sheet), and $\Delta\varepsilon$ is the change in strain due to applied stress.

Table 7.3 summarizes the maximum stress at failure obtained using this approach for different crack length cases, exhibiting a clear crack length dependence.

TABLE 7.3
Maximum Traction Obtained for Graphene Sheet Using the Change in Overall Potential Energy Method from MD

Crack Length (Å)	Change in Potential Energy ($\Delta\pi$) (J)	Strain at Failure	Max Traction (σ_{max}) (GPa)
0	1.88×10^{-15}	0.15	119.05
2.5	9.40×10^{-16}	0.10	89.39
7.6	3.50×10^{-16}	0.06	55.62
17.7	1.70×10^{-16}	0.04	38.24

Figure 7.10 depicts the traction-separation-law obtained for a graphene sheet with half-crack lengths of 0, 7.6, and 17.7 Å, respectively, using Equation (7.6).

Comparison of Figure 7.10 with Figures 7.7 and 7.8, respectively, indicates that the potential energy approach does not provide accurate data for the near-tip traction-separation law due to the global nature of the potential energy metric. Further, comparison of Table 7.3 with Table 7.1 reveals that the potential energy approach agrees reasonably with far-field traction-separation behavior but suffers from a similar crack-length dependence pathology. It should be noted that Figure 7.3 provides traction-separation law obtained from far-field stress-strain data, whereas

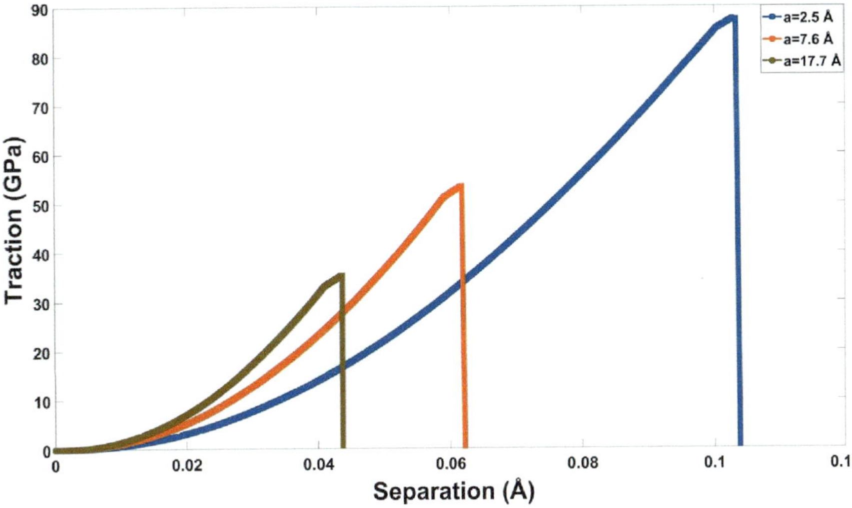

FIGURE 7.10 The graph depicts the traction varying with to respect to far field separation for different half-crack lengths obtained from changes in potential energy of the MD system.

Figure 7.10 provides traction-separation law that was obtained from change in potential energy of the system. The difference in convexity of the curves between Figures 7.3 and 7.10 is likely due to the different energy-based approaches used, but they both indicate crack-length dependence of the traction-separation law and therefore are not reliable.

7.3 FRACTURE MODELING USING CONCURRENT MD-FEM COUPLING

Based on the techniques introduced in Chapter 6 related to concurrent coupling of FEM with MD, simulation results are presented in this section for concurrent coupling applied to predicting fracture in monolayer graphene.

7.3.1 Fracture Modeling in Monolayer Graphene Using Concurrent MD-FEM Coupling

A schematic of the coupled FEM and MD domains, along with a bridging domain, is shown in Figure 7.11.

The MD domain is 30 × 30 nm in size, the FEM domain is 90 × 90 nm in size, and the bridging domain is 4 nm wide. A total crack length of $2a = 8$ nm was created

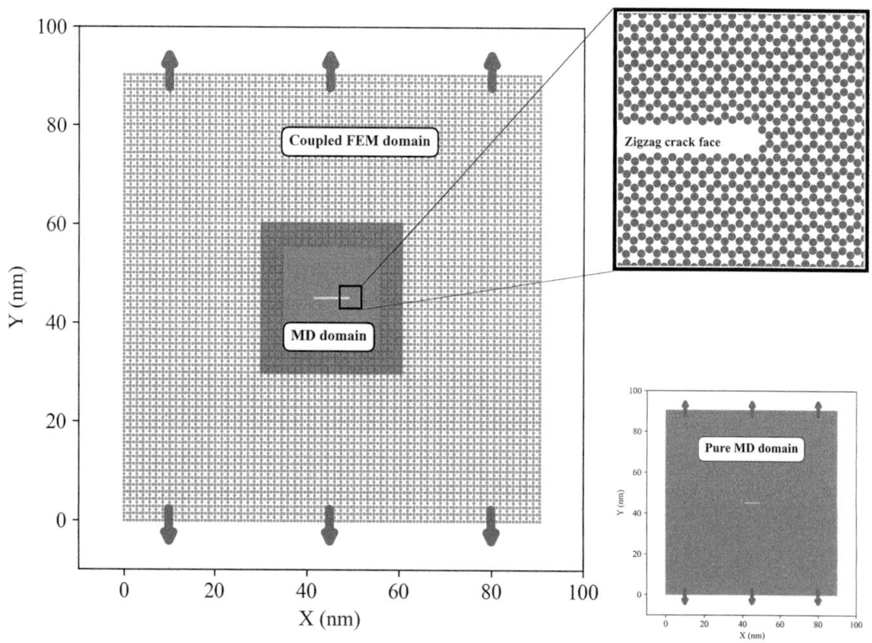

FIGURE 7.11 Schematic of coupled FEM-MD fracture simulation on monolayer graphene at $T = 1$ K and $T = 300$ K. The chirality along the crack face is "zig-zag".

by deleting atoms. Far-field uniaxial tensile loading is applied to the system in the Y direction as shown in Figure 7.11. Loading is applied in strain increments of 2% and 0.5% for simulation cases of $T = 1$ K and $T = 300$ K, respectively, as finer increments are required for stable simulations at higher temperatures. Details on the iterative coupling between MD and FEM can be found in Chapter 6, with the "Direct" coupling approach being used in the present problem. For comparison with the results of the coupled MD simulations, pure MD simulations for monolayer graphene were also performed at $T = 1$ K and $T = 300$ K. The size of the pure MD specimens was the same as that of the total coupled simulation specimens, i.e., 90 × 90 nm. In the pure MD specimens, loading is applied at a constant displacement rate to the top and bottom domain edges, resulting in a strain rate of 5 × 10^8/s. For the coupled simulations, 500 iterations were performed at each load increment to ensure proper convergence of energy, forces, and displacements in the coupled domains, with each MD iteration performed over 20 ps of simulation time.

The energetic component of the atomistic J-integral (J_U) was computed for both the coupled simulations and pure MD simulations. The results of the computed J-integral for the $T = 1$ K case are plotted with respect to the applied far-field strain in Figure 7.12.

The red line shows the J-integral computed from the pure MD simulation, which increases smoothly due to the constant rate of applied displacement. On the other

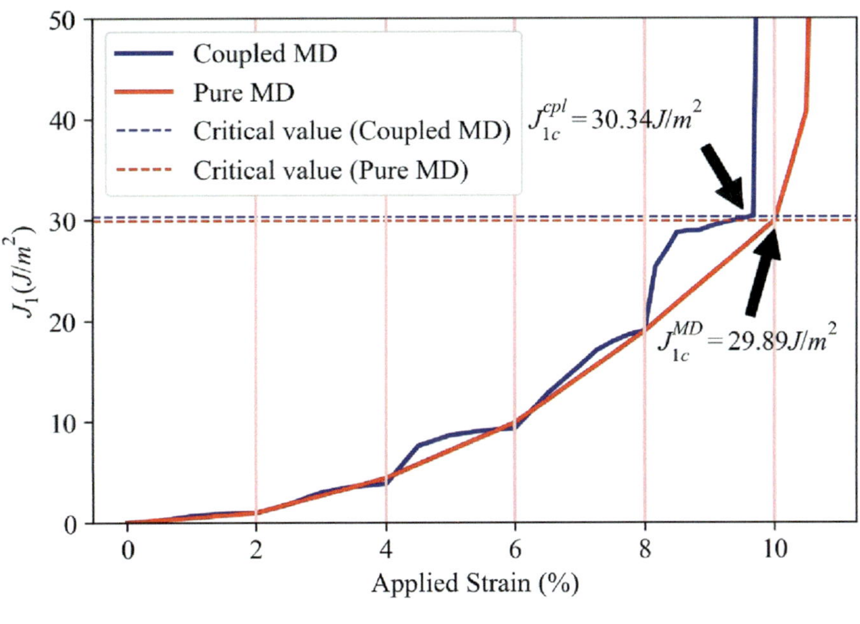

FIGURE 7.12 J-integral variation with applied far-field strain for coupled MD-FEM fracture simulation on monolayer graphene at $T = 1$ K.

hand, the blue line shows the results from the MD-FEM coupled simulation, which exhibits more complex patterns. At each level of applied loading, the coupled MD-FEM computed J-integral value initially overshoots the pure MD data, and then slowly converges to the correct value as the coupled iteration progresses and minimizes the error, as can be seen in Figure 7.12. Finally, at the end of 500 coupling iterations, the blue curve appears to meet closely with the red curve, i.e., the converged J-integral values match well with the pure MD J-integral. A similar trend is observed in the $T = 300$ K case (not shown here), though it is noisier due to thermal fluctuations. Finally, at around 10% strain, crack initiation occurs as can be seen by the spike in the J-integral value in Figure 7.12 for pure MD as well as MD-FEM coupling, indicating the onset of brittle fracture. The spike occurs due to the release of large amount of stored energy at the crack tip when the graphene bond breaks, and this energy release rate is correctly captured by the J-integral, followed by rapid crack growth. The errors in the estimated fracture toughness J_{1c} at $T = 1$ K and $T = 300$ K when comparing the coupled results with the pure MD results are approximately 2% and 12%, respectively, indicating good overall coupling between FEM and MD. The stress distributions immediately prior to damage occurring in the coupled and pure MD cases at $T = 300$ K, along with the relative error distribution, are shown in Figure 7.13.

As expected, the highest error in stress is near the corners of the bridging domain, as was discussed in Chapter 6. The error in the stress at the crack tips comparing coupled and pure MD cases is approximately 5%.

Finally, in Figure 7.14, the visualizations of self-similar crack growth in both coupled MD and pure MD simulations are shown at a comparable loading level. The exact point at which the crack initiates, as well as the nature of crack growth, is very similar in the two cases. In the coupled MD case, the crack does not grow further as it has impinged on the bridging domain boundary.

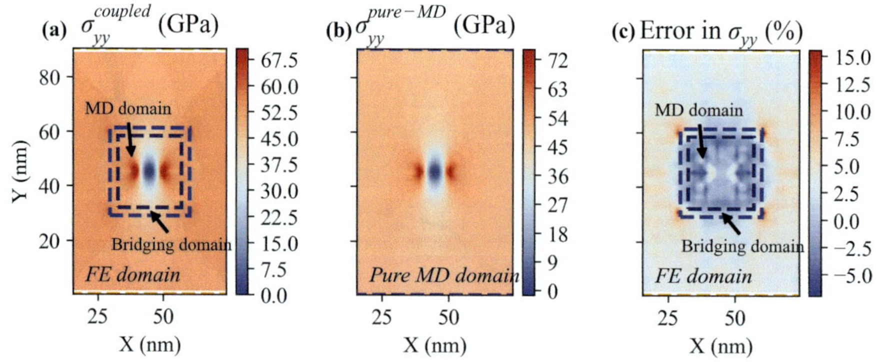

FIGURE 7.13 Stress σ_{yy} contour plots for coupled MD-FEM (left), and pure MD (center), in fracture simulations at $T = 300$ K of a center-cracked monolayer graphene sheet, along with relative error plot (right).

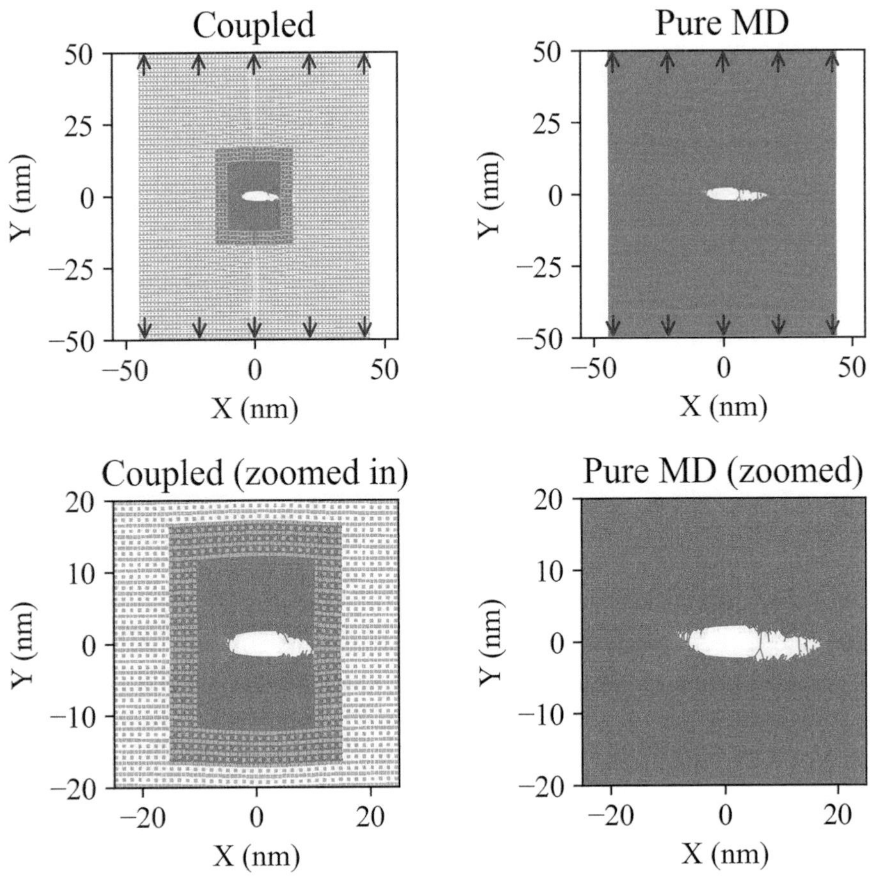

FIGURE 7.14 Visualization of brittle self-similar crack growth in monolayer graphene at $T = 300$ K_coupled simulation (left) and pure MD simulation (right) at $T = 1$ K.

7.4 SUMMARY

In addition to an illustrative case study of concurrent coupling being employed to study fracture in monolayer graphene described in Section 7.3, an important concern regarding fracture simulations using cohesive zone models at the atomistic scale was discussed in this chapter. There are two primary issues with the traditional practice of employing the far-field traction-separation data from MD simulations to obtain the near-tip traction-separation law at the nanoscale: (a) the far-field traction-separation data, usually obtained using MD or Finite Element Analysis (FEA), does not correctly represent the actual shape of the near-tip traction-separation law, and (b) the far-field stress-strain data shows a steady decrease in failure stress with increasing crack length, and therefore is not a true material property. The reason for this anomalous behavior was described in rigorous detail in Section 7.2 through the use of the atomistic J-integral, and the correct J-integral contour for obtaining a traction-separation law that is independent of crack size was clearly demonstrated. Again, the key requirement here is that the fully developed fracture process zone must remain within the narrow J-contour, which is satisfied in the case of monolayer graphene with a fully developed FPZ size of 0.08 nm as listed in Table 5.1.

REFERENCES

1. Zhang, P., et al., Fracture toughness of graphene. *Nature Communications*, 2014. **5**(1): p. 3782.
2. Zhou, X.W., J.A. Zimmerman, E.D. Reedy Jr, and N.R. Moody, Molecular dynamics simulation based cohesive surface representation of mixed mode fracture. *Mechanics of Materials*, 2008. **40**: p. 832–845.
3. Kai, H., T. Shimada, N.I. Ozaki, Y. Hagiwara, T. Sumigawa, L. Guo, and T. Kitamura, A unified and universal Griffith-based criterion for brittle fracture. *International Journal of Solids and Structures*, 2017. **128**: p. 67–72.
4. Dewapriya, M.A.N., R.K.N.D. Rajapakse, and A.S. Phani, Atomistic and continuum modelling of temperature-dependent fracture of graphene. *International Journal of Fracture*, 2014. **187**: p. 199–212.
5. Subramanian, N., A. Rai, and A. Chattopadhyay, Atomistically derived cohesive behavior of interphases in carbon fiber reinforced CNT nanocomposites. *Carbon*, 2017. **117**: p. 55–64.
6. Awasthi, A.P., D.C. Lagoudas, and D.C. Hammerand, Modeling of graphene–polymer interfacial mechanical behavior using molecular dynamics. *Modelling and Simulation in Materials Science and Engineering*, 2009. **17**: p. 015002 (37pp).
7. Krull, H. and H. Yuan, Suggestions to the cohesive traction–separation law from atomistic simulations. *Engineering Fracture Mechanics*, 2011. **78**: p. 525–533.
8. Dandekar, C.R. and C. Shin Yung, Molecular dynamics based cohesive zone law for describing Al–SiC interface mechanics. *Composites: Part A*, 2011. **42**: p. 355–363.
9. Roy, S. and A. Roy, A computational investigation of length-scale effects in the fracture behavior of a graphene sheet using the atomistic J-integral. *Engineering Fracture Mechanics*, 2019. **207**: p. 165–180.
10. Jones, R.E., J.A. Zimmerman, J. Oswald, and T. Belytschko, An atomistic J-integral at finite temperature based on Hardy estimates of continuum fields. *Journal of Physics: Condensed Matter*, 2010. **23**(1): p. 015002.
11. Zhang, T., X. Li, S. Kadkhodaei, and H. Gao, Flaw insensitive fracture in nanocrystalline graphene. *Nano Letters*, 2012. **12**: p. 4605–4610.
12. Stuart, S.J., A.B. Tutein, and J.A. Harrison, A reactive potential for hydrocarbons with intermolecular interactions. *The Journal of Chemical Physics*, 2000. **112**: p. 6472.
13. Chowdhury, S.C. and T. Okabe, Computer simulation of carbon nanotube pull-out from polymer by the molecular dynamics method. *Composites: Part A*, 2007. **38**: p. 747–754.
14. Gao, H. and S. Chen, Flaw tolerance in a thin strip under tension. *The Journal of Applied Mechanics*, 2005. **72**: p. 732–737.

8 Modeling of Fracture in Amorphous Materials Using MD

8.1 INTRODUCTION

The focus of this chapter is the application of the "K-test" approach developed in previous chapters along with the atomistic J-integral to modeling and predicting fracture toughness in amorphous materials. Two amorphous materials, both existing in glassy states at room temperature, are studied in this chapter. The first material is epoxy resin, which is a thermoset polymer and is an important matrix material used in fiber-reinforced polymer (FRP) composites [1]. Failure modes in FRPs typically relate to either matrix fracture or fiber/matrix debonding. The fracture toughness of the epoxy matrix can be enhanced by embedding nanoparticles such as silica [2–4], rubber [5, 6], alumina [7, 8], carbon nanotubes (CNT), and graphene nanoplatelets (GNP) [9–11]. Accelerating the inverse design of polymer nanocomposites with superior fracture toughness using first-principles simulations necessitates the development of reliable MD simulation techniques to predict fracture toughness. The second material that is investigated in this chapter is amorphous carbon, which has applications in thin film coatings for electronics and solar cells [12, 13].

As will be explained in detail in this chapter, the fully developed FPZ size of crosslinked epoxy is relatively large (11,440 nm) compared with tractable MD simulation domain size. On the other hand, amorphous carbon offers an intermediate fully developed FPZ length scale (14 nm), which is large enough to observe and study the nature of fracture, but not too large that fracture simulations are computationally intractable using MD. Firstly, the modeling approach and fracture results for epoxy are described, including a careful examination of stress and yielding near the crack tip. Following this, a study of fracture in amorphous carbon is presented with actual quantitative fracture toughness predictions and validation against experimental data in the literature.

8.2 MD FRACTURE MODELING IN CROSS-LINKED EPOXY POLYMER

In previous chapters, such as Chapter 5, some selected results on MD simulations of epoxy polymers were presented. In this chapter, a more detailed description of the MD modeling of amorphous epoxy polymers is provided, following which fracture simulation results are discussed.

DOI: 10.1201/9781003440734-8

8.2.1 MD Simulation Methods for Modeling Epoxy

The epoxy polymer that is considered here consists of EPON862 monomers (Diglycidyl Ether of Bisphenol F) with a crosslinking agent Diethyl toluene diamine (DETDA). Atomistic modeling is performed using the OPLS potential [14–16], which is a fixed-bond potential consisting of harmonic bond, angle, and dihedral energy terms as was described in Chapter 3. The technique to crosslink epoxy polymers based on fixed bond potentials is now well established, with simulation models exhibiting good agreement with experimental data on various thermo-mechanical properties [17, 18]. The crosslinking procedure involves *a priori* specification of the necessary reaction pathways, and programmatic formation and deletion of bonds. Typically, crosslinking is performed above the glass transition temperature (T_g), followed by a curing phase. After curing, the harmonic bonds of the OPLS potential are replaced by Morse bonds, which allow bond breakage, as shown in Figure 8.1 based on Morse parameters calculated in [19].

It should be noted that harmonic bond potentials, such as OPLS, cannot accurately model bond breakage due to fracture, and hence need to be replaced by an atomic potential that correctly accounts for bond separation, such as the Morse potential after epoxy crosslinking has been performed.

The MD simulations presented in this chapter were performed on 86% crosslinked epoxy resin. Prior to performing fracture simulations, the elastic-plastic properties of

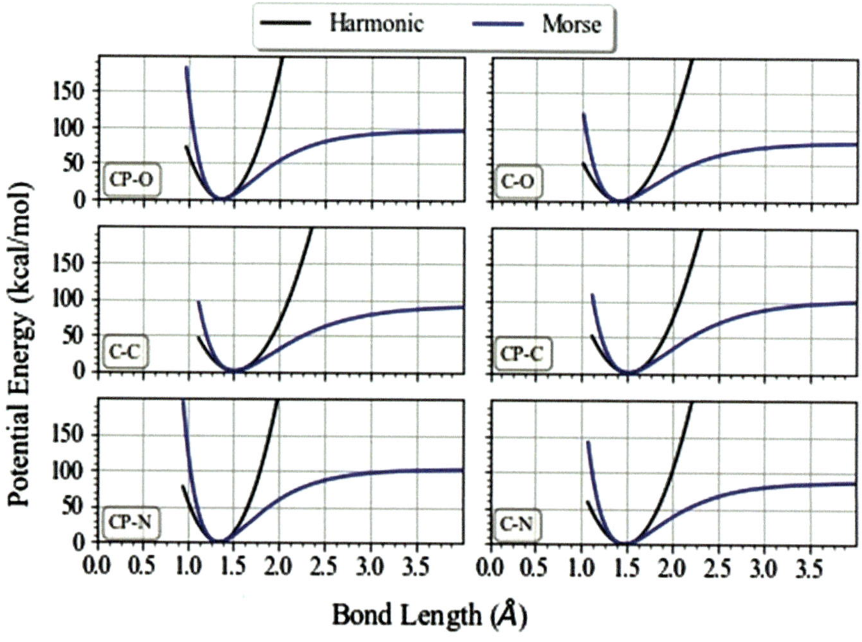

FIGURE 8.1 Comparison of harmonic bond potentials in epoxy polymer from OPLS potential with corresponding Morse potential replacements to enable bond breakage at critical bond length.

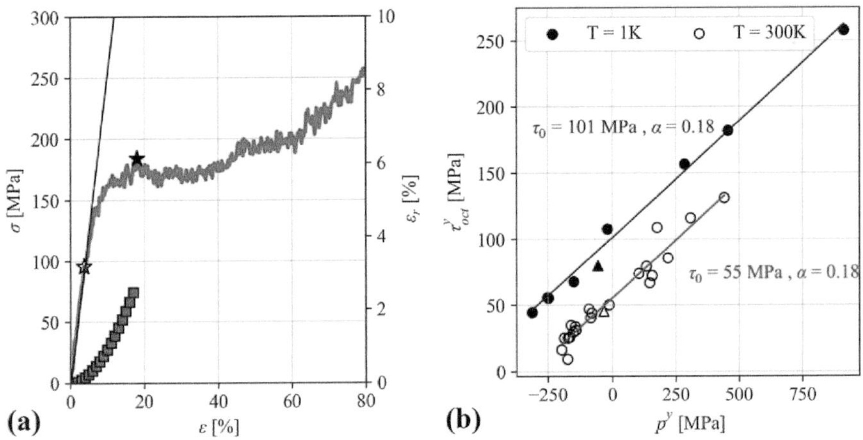

FIGURE 8.2 (a) Stress-strain relation from MD-simulated uniaxial tension tests on 86% crosslinked epoxy at $T = 300\,\mathrm{K}$, along with residual strain plotted based on unloading simulations, (b) octahedral stress at yield plotted against pressure at yield for various triaxial loading cases and temperatures.

the crosslinked MD model were interrogated using a representative volume element (RVE) without any cracks. Simulated uniaxial tension tests were performed in MD, for both loading and unloading of epoxy, using the OPLS-Morse potential. The uniaxial stress-strain relation for epoxy obtained from the simulation is shown in Figure 8.2a for up to 80% strain (red curve), indicating an elastic-strain-hardening-plastic behavior at $T = 300\,\mathrm{K}$.

Also shown in Figure 8.2a is the residual inelastic strain (black symbols) based on the unloading data. Because the yield strength of polymers is dependent on the state of triaxial stress, a series of triaxial test simulations were also performed in MD, and octahedral stress and hydrostatic pressure at yield were recorded similar to techniques in [20, 21]. These results are shown in Figure 8.2b and fitted to yield a pressure-modified von-Mises (PMvM) yield criterion [22] for epoxy:

$$\tau^y_{\mathrm{oct}} = \tau_0 + \alpha p^y \tag{8.1}$$

where τ^y_{oct} denotes the octahedral stress at yield, p^y denotes the pressure at yield, and τ_0 and α are coefficients of the assumed linear relationship. The values for τ_0 and α obtained via linear regression fit are shown in Figure 8.2b for $T = 1$ and 300 K respectively. The Young's modulus at $T = 300\,\mathrm{K}$ and glass transition temperature predicted by our model agree very well with experimental data from the literature [23–26]. It should be noted that macroscale experiments on epoxy resin typically cannot capture the inelastic constitutive behavior up to 80% strain as seen in MD simulations, due to defects and voids that act as stress concentrators in test specimens in macroscale tension tests. However, experimental compression tests exhibit very similar stress-strain behavior, with a hardening-softening-hardening curve [27, 28] similar to that seen in our MD simulations. Moreover, our MD results largely agree with other MD studies in the literature with regard to Young's modulus, yield stress, and glass transition temperature for epoxy [18, 29]. Young's modulus values predicted by our MD model are 2.6 GPa at room temperature ($T = 300\,\mathrm{K}$) and 5.2 GPa at $T = 1$ K, respectively, for epoxy with 86% crosslink density. Similarly, yield stress values predicted by our model are 95 MPa at $T = 300\,\mathrm{K}$, and 169 MPa at 1 K, respectively. The glass transition temperature for epoxy predicted by our model is $T_g = 416$ K (i.e., 143°C).

8.2.2 *K*-Test Fracture Simulation Results for Epoxy Resin

To perform *K*-test fracture simulations on epoxy resin, a MD domain geometry such as the one shown in Figure 8.3 was constructed.

The *X* and *Y* dimensions of the domain are significantly larger than the out-of-plane (*Z*-direction) thickness. Moreover, periodic boundary conditions are used in the *Z* direction, therefore representing a PS/PE model. The MD specimen size is approximately 44 × 44 × 6 nm with a 10 nm pre-crack created by deleting atoms and also disabling interatomic interactions across the crack face. The effective crack length in the *K*-test approach, however, as described in Chapter 5, is much larger as the application of the *K*-test boundary condition assumes the existence of a singular elastic stress field near the crack tip, outside the FPZ, and hence full notch sensitivity. In an attempt to study the FPZ in detail, the simulation domain size used in our work with ~1.1 million atoms is significantly larger than most simulations typically performed in the literature. The *K*-field loading was applied to the epoxy polymer specimens using SMD in increments of $\Delta K_1 = 0.01\ \mathrm{MPa}\sqrt{\mathrm{m}}$ with equilibration times of 100 ps after each loading step. Simulations were performed at $T = 1$ K and $T = 300\,\mathrm{K}$, as well as using PS and PE conditions. As expected, the stress fields

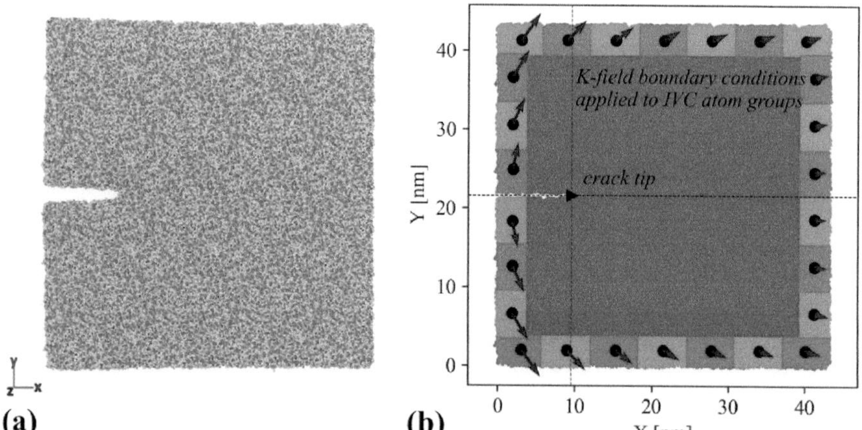

FIGURE 8.3 *K*-test fracture simulation geometry for epoxy resin (a) atomistic visualization, (b) showing *K*-test boundary groups and *K*-field displacements applied at the boundary.

in the higher-temperature simulations ($T = 300$ K) exhibit significant noise due to the random thermal motion of atoms and are more challenging to study. In Figure 8.4, the stress components (von-Mises and volumetric stress) ahead of the crack tip at $T = 1$ K and under PE conditions are plotted along with LEFM-predicted stress based on asymptotic stress fields presented in Chapter 5. The MD-predicted stress results show excellent agreement with LEFM results away from the crack tip. Very close to the crack tip, the material begins to yield, as displayed in Figure 8.4 with red symbols, based on the pressure-modified von-Mises (PMvM) yield criterion.

Interestingly, besides plastic yielding, the formation of nanoscale voids in the FPZ is also captured as noted by the significant dips in the stress profile at higher applied K values. It should be noted that three independent realizations of the MD K-test specimen were crosslinked and tested to capture stochastic variations in the data, as indicated by the grey envelope in Figures 8.4 and 8.5. Also, the sharp decline in the stresses at a distance of 40 nm from the crack tip is due to the presence of the IVC atomic groups located at the K-test boundary, and not due to voids.

In Figure 8.5, similar plots are shown for simulations at $T = 300$ K and under PE conditions. As alluded to earlier, the results are significantly noisier compared with results at $T = 1$ K due to thermal fluctuations of polymer molecules.

However, qualitatively, the trend is similar in both cases. Notable differences include the fact that significant yielding occurs at lower applied loading at $T = 300$ K compared with $T = 1$ K. This is due to the fact that the yield stress at $T = 300$ K is lower than that at $T = 1$ K, because of which the FPZ radius, given by Equations (8.2a) and (8.2b) for PE and PS respectively, is larger based on Irwin's

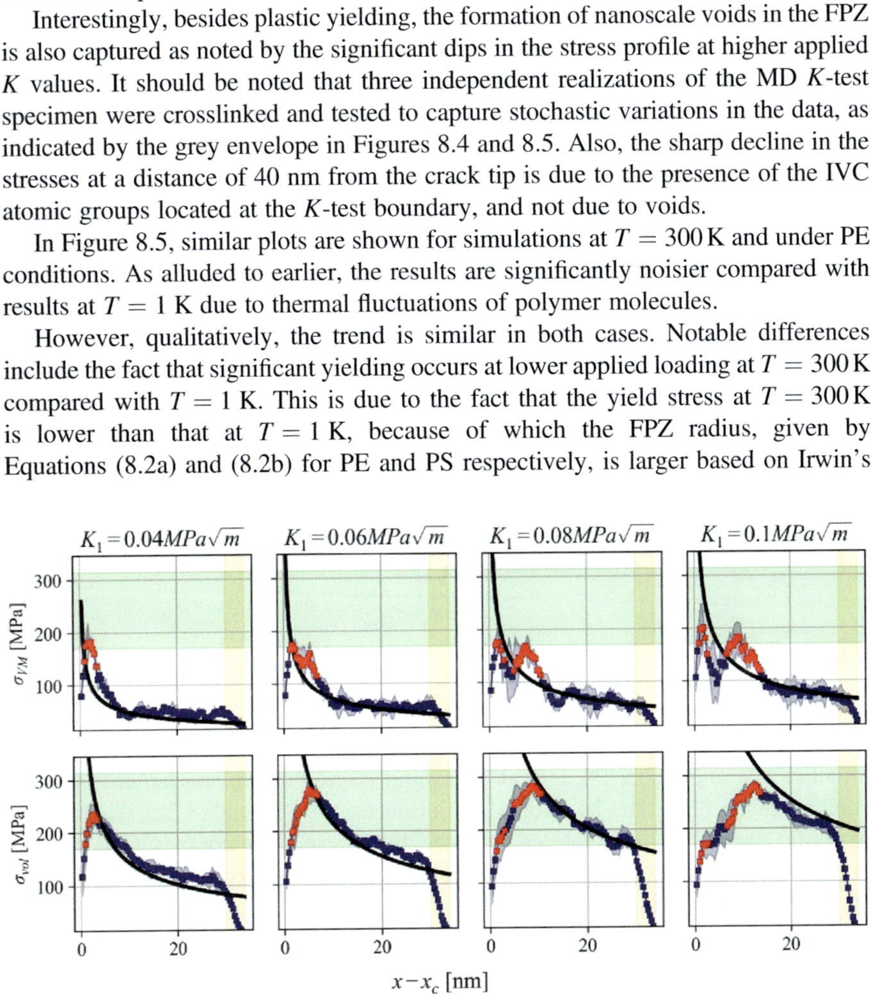

FIGURE 8.4 MD prediction of von-Mises (top row) and volumetric (bottom row) stress as a function of distance from the crack tip at increasing load levels at $T = 1$ K under PE conditions, plotted alongside asymptotic LEFM stress (black solid lines) for comparison. Red color signifies yielding based on the PMvM yield criterion, and the grey envelope indicates the standard deviation of the MD data.

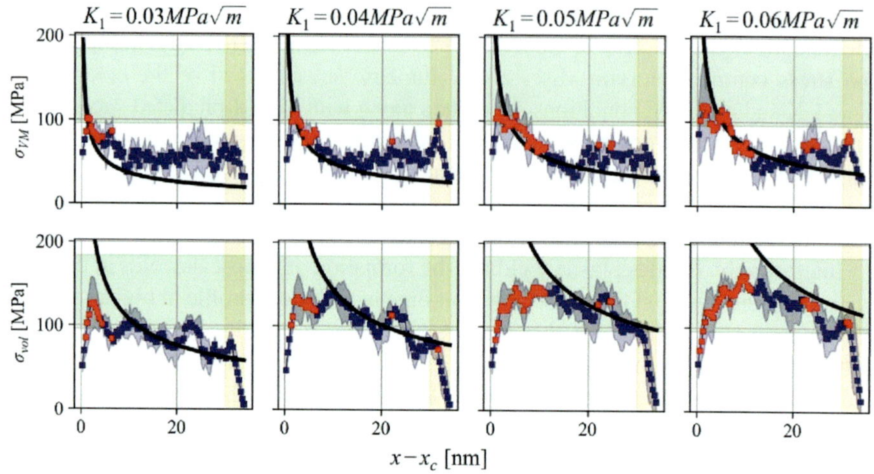

FIGURE 8.5 MD prediction of von-Mises (top row) and volumetric (bottom row) stress ahead of the crack tip at increasing load levels at $T = 300\,\text{K}$ under PE conditions, plotted alongside asymptotic LEFM stress for comparison (black solid lines). Red symbols signify yielding based on the PMvM yield criterion, and the grey envelope indicates the standard deviation of the MD data.

plastic zone model [30], where σ_{yld} is the yield stress and K_1 is the Mode-I stress intensity factor:

$$r_{\text{FPZ}}^{\text{PE}} = \frac{1}{3\pi}\left(\frac{K_1}{\sigma_{\text{yld}}}\right)^2 \tag{8.2a}$$

$$r_{\text{FPZ}}^{\text{PS}} = \frac{1}{\pi}\left(\frac{K_1}{\sigma_{\text{yld}}}\right)^2 \tag{8.2b}$$

Simulation results under PS conditions exhibit more yielding and larger FPZ sizes compared with PE, as can be inferred from Equations (8.2a) and (8.2b), due to a loss of elastic constraint. In the interest of space, those results are not presented in detail here. A comparison of von-Mises stress ahead of the crack tip under PE and PS conditions is shown in Figure 8.6a.

Clearly, the von-Mises stress level is higher under PS conditions due to a more developed yield zone, with significant deviation from LEFM at the given loading level. The green band in Figure 8.6a indicates the yield stress range. Figure 8.6b shows the atomistic J-integral, computed on contours surrounding the crack tip, plotted as a function of the applied Mode-I stress intensity factor K_1. Both PE and PS cases show excellent agreement with their corresponding LEFM curves at low loading levels, i.e., prior to significant FPZ growth. Eventually, the MD data in both cases deviate from the LEFM curves, which continue to rise. The reason for this deviation is the fact that the FPZ has grown so large that small-scale yielding (SSY)

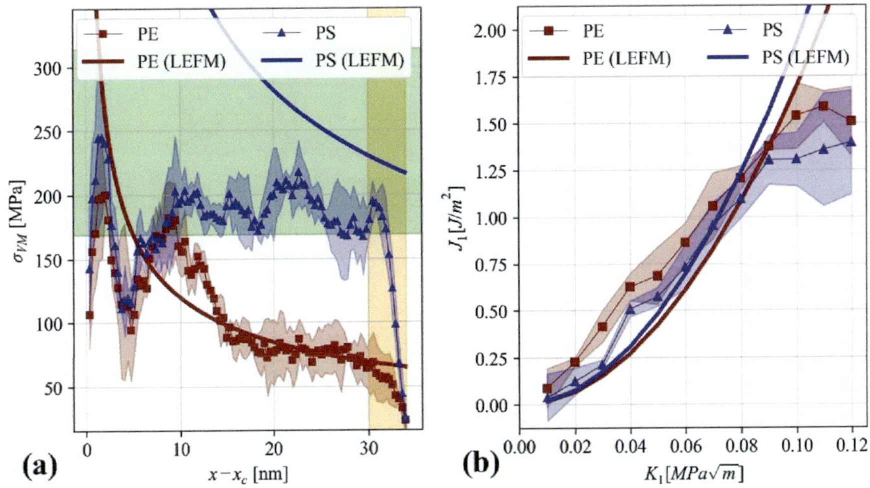

FIGURE 8.6 Comparison of K-test MD fracture results for epoxy under PE and PS conditions at $T = 1$ K: (a) von-Mises stress ahead of the crack tip at $K_1 = 0.1$ MPa$\sqrt{\text{m}}$, (b) atomistic J-integral vs. K_1 data for epoxy at $T = 1$ K. The green band in (a) indicates the yield stress range.

conditions are no longer satisfied; therefore, the K-test boundary conditions are no longer applicable. Therefore, despite using the K-test approach and relatively large MD domain size (compared with literature), the fully developed FPZ for epoxy resin was not captured in this work, and therefore, brittle fracture could not be demonstrated. A combination of coarse-graining and concurrent coupling is expected to alleviate the situation. However, significant improvements in computing power in the future will be required, regardless of these techniques, to enable the prediction of brittle fracture in glassy polymers with a fully developed FPZ. It should be noted that the authors have attempted to perform concurrent FEM-MD coupling simulations to predict fracture in epoxy and published their results in [31]. However, as the FPZ size grows with increasing loading, the continuum representation of the material becomes invalid well before the fully developed FPZ is achieved. This is because the continuum FEM domain surrounding the MD domain into which the FPZ propagates with increasing load is not capable of accurately modeling the void and fibril formation that MD is capable of modeling. This is evidenced in Figure 8.7a by the formation of voids and fibrils, a phenomenon termed crazing in polymers, at around 10% strain near the crack tip.

The resulting reduction in volumetric stress, due to the loss of triaxial constraint near the crack tip, is evident in the volumetric stress plot shown in Figure 8.7b. Consequently, predicting fracture toughness of epoxy polymer currently remains beyond the reach of computational fracture mechanics primarily due to the characteristic length scale issues highlighted in Chapter 5.

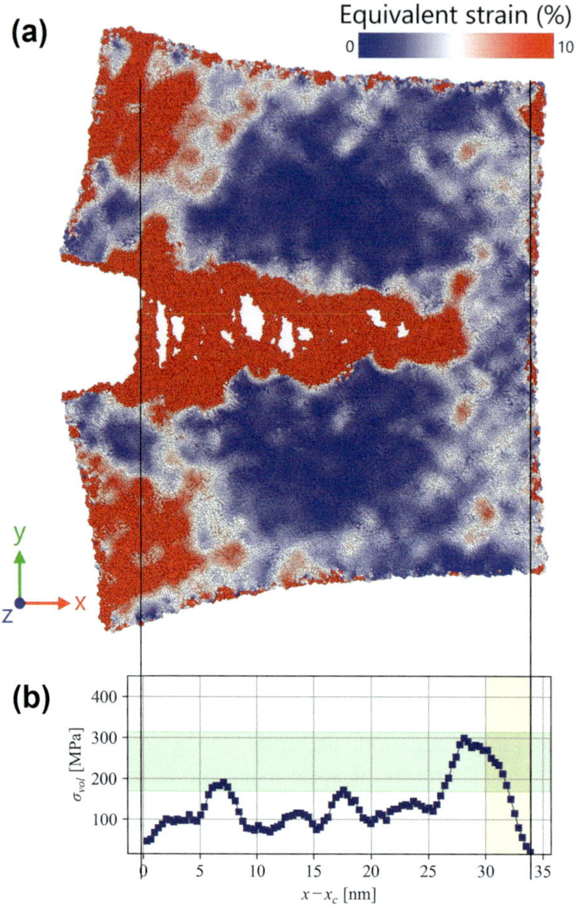

FIGURE 8.7 (a) Breakdown of epoxy continuum at the nanoscale as evidenced by formation of voids and fibrils (b) corresponding reduction of volumetric stress ahead of crack tip.

8.3 MD FRACTURE MODELING IN AMORPHOUS CARBON

Due to the inherent challenges in modeling brittle fracture in glassy polymers using MD simulations, namely due to FPZ length scale issues, in this section we shift our focus to amorphous carbon. As noted in the literature [32], amorphous carbon has a fully developed FPZ size that is computationally in the "sweet spot," i.e., large enough to study interesting nonlinear, inelastic phenomena near the crack tip, while remaining computationally tractable. In this section, we apply similar techniques as were applied to epoxy to the modeling of fracture in amorphous carbon (a-C).

8.3.1 MD Modeling Methods for Amorphous Carbon

In this section, amorphous carbon is modeled using the Tersoff potential [33], which is an efficient, bond-order-based potential capable of bond breakage. Synthesis was performed by a melt-quench procedure [32, 34] at $T = 6000$ K and a cooling rate of

TABLE 8.1
Summary of Thermo-Mechanical Properties of Amorphous Carbon Simulated Using Tersoff Potential in LAMMPS at $T = 300\,K$

Property	Value
Young's modulus	450 ± 5 GPa
Poisson's ratio	0.12
0.2% yield stress	22 ± 1 GPa
Peak stress	86 ± 1 GPa
MD strain rate	5×10^8/s

1000 K/ps with density 2.5 g/cc. Uniaxial tension tests were performed on representative volume elements (RVE) to predict the bulk properties of the simulated amorphous carbon material. A summary of key thermo-mechanical properties based on MD simulations at $T = 300\,K$ are listed in Table 8.1. The literature on the experimental characterization of amorphous carbon exhibits significant variance [35–37] with Young's modulus values ranging from 100 to 500 GPa for specimens with mixed sp2/sp3 bonding and greater than 700 GPa for specimens derived from diamond with very high sp3 bonding. Properties also depend on the density of amorphous carbon. Considering this variability, our simulation model is deemed adequate. Moreover, our predicted properties agree well with previous MD studies in the literature [32, 38]. The applied MD strain rate is 5×10^8/s.

8.3.2 *K*-Test Fracture Toughness Predictions in Amorphous Carbon

Due to the simplicity in atomistic structure and homogeneity of amorphous carbon relative to epoxy polymer, the thermal fluctuations are easier to handle in this case. Therefore, all fracture simulations were performed at $T = 300\,K$ without significant noise in the results. K-test simulations were performed on three different specimen sizes (varying in-plane dimensions X and Y) as shown in Figure 8.8, in order to study specimen size effect on predicted toughness.

The out-of-plane dimension (Z) was relatively thin (2.5 nm) with periodic boundary conditions, as was done for epoxy. PS conditions were used, though significant out-of-plane strain was not observed due to the relatively small Poisson's ratio of the atomistic model. K-test loading was applied in increments of $\Delta K_1 = 0.1\ \text{MPa}\sqrt{\text{m}}$. The applied MD strain rate was 5×10^8/s.

Unlike epoxy polymer, due to the moderate size of the fully developed FPZ for amorphous carbon (~12 nm), the quantitative prediction of fracture toughness in amorphous carbon was possible. Figure 8.8 shows the fully developed FPZ, immediately prior to crack initiation. Fracture process zone (FPZ) contours are drawn based on the von-Mises stress and the predicted yield stress of the material (refer to Table 8.1). The MD-predicted contours show excellent agreement with the

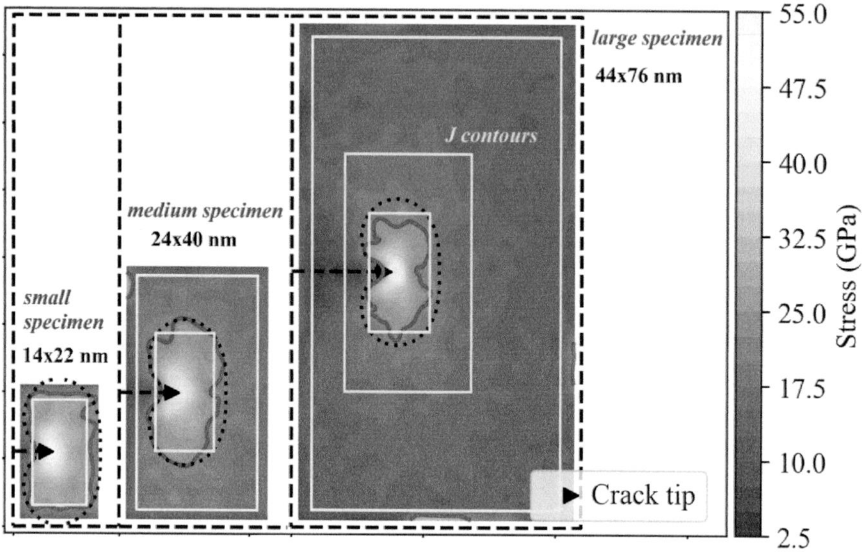

FIGURE 8.8 K-test MD simulations performed on three different specimen sizes, 14 × 22 nm, 24 × 40 nm, and 44 × 76 nm at $T = 300$ K. Fully developed FPZ sizes are shown via von-Mises stress contour plots, as predicted by MD (red solid contour) and LEFM (black dotted contour). The three concentric J-integral contours used are displayed in yellow. Black arrow indicates crack tip location.

LEFM-predicted contours, thereby reinforcing the validity of the results. The smallest specimen size (14 × 22 nm) appears to be insufficiently large to completely capture the fully developed FPZ of 12 nm and therefore, results are not reliable. The other specimens (24 × 40 nm and 44 × 76 nm) are much closer to the small-scale yielding (SSY) assumptions which are a part of the K-test approach. The three concentric J-integral contours used in the three specimen sizes are shown using yellow lines. Clearly, the smallest contour is not large enough to capture the fully developed FPZ. By examining the stress distribution ahead of the crack tip immediately prior to crack initiation as a log-log plot shown in Figure 8.9, we can formally assess the validity of our approach.

A clear line of demarcation, indicated by the vertical black line in Figure 8.9, is observed between the fracture process zone (FPZ) and the K-dominance zone. The plot clearly shows that within the FPZ the LEFM-based assumption of $1/r^{0.5}$ singularity is no longer valid as the MD-predicted slope of the stress vs. distance from the crack tip dataset is greater than -0.5. As a final validation of the K-test approach in quantitatively predicting fracture toughness in amorphous carbon, we study the atomistic J-integral plotted as a function of K_I as shown in Figure 8.10.

The results from all three K-test specimen sizes used are shown alongside LEFM predictions for PS. At low K_I, all three specimen sizes show excellent agreement with LEFM. As the FPZ develops at higher loading, the smallest specimen (shown in blue) begins to underpredict, which is the result of the smaller J-integral contour becoming overtaken by the growing FPZ as shown in Figure 8.8, and hence losing its

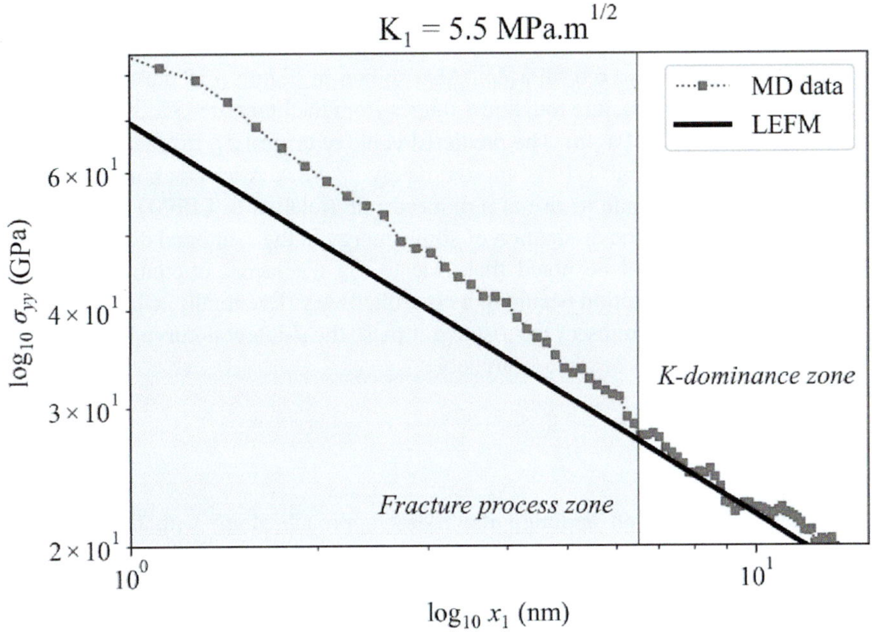

FIGURE 8.9 Log-log plot of normal stress (in red), as predicted by MD simulations, as a function of distance ahead of the crack tip, immediately prior to crack initiation at $K_1 = 5.5\,\text{MPa}\sqrt{\text{m}}$. LEFM data is plotted for comparison and the MD data shows clear deviation in the FPZ.

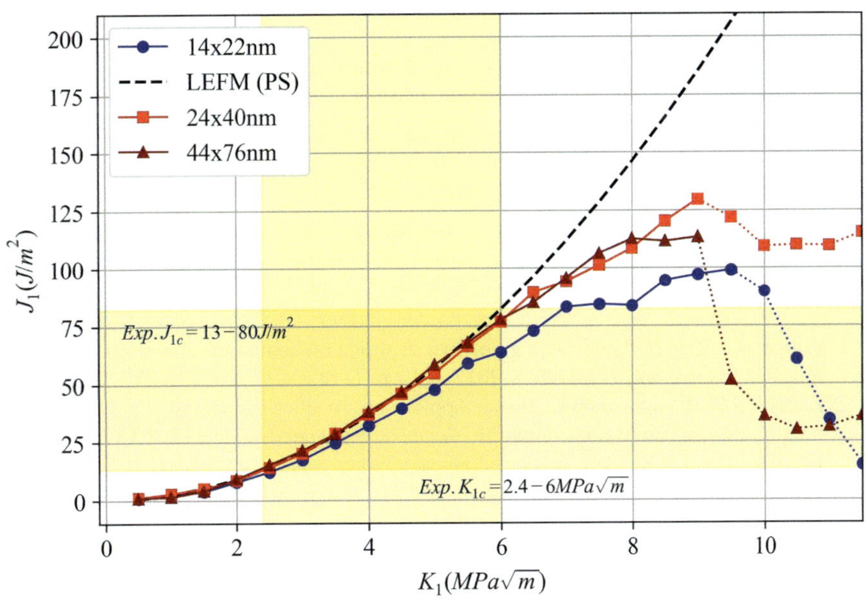

FIGURE 8.10 Atomistic J-integral plotted as a function of Mode-I stress intensity factor K_1 for amorphous carbon at $T = 300\,\text{K}$ with results from three K-test specimen sizes shown alongside the LEFM curve (black dashed line).

path independence attribute. On the other hand, the larger K-test specimens (shown in red and burgundy) exhibit almost perfect agreement with LEFM up until crack initiation occurs at $K_\mathrm{I} = 6.0\ \mathrm{MPa}\sqrt{\mathrm{m}}$. Also shown in Figure 8.10 are the available experimental Mode-I fracture toughness ranges from the literature [32, 39–41] which are $K_{1C} = 2.4 - 6.0\ \mathrm{MPa}\sqrt{\mathrm{m}}$. The predicted value of critical J_{1c} evaluated from our MD simulations in Figure 8.10 is approximately 78 J/m^2. After crack initiation, the J-integral curves continue to rise at a reduced rate (relative to LEFM) as the crack propagates, caused by the dissipation of strain energy in the J-integral domain due to crack growth. It should be noted that, due to the displacement-controlled K-test approach, crack propagation occurs in a controlled way. Eventually, when the crack grows long enough to transect the J-integral path, the J-integral curve begins to go down and the MD data becomes unreliable.

8.4 SUMMARY

This chapter focused on applying the K-test approach along with the atomistic J-integral to simulate fracture in crosslinked epoxy resin and amorphous carbon. Stress distributions and the evolution of the fracture process zone (FPZ) were observed for epoxy at low and high temperatures under PE and PS conditions. For epoxy polymer, brittle crack growth could not be simulated due to length scale constraints. Instead, a breakdown of the continuum behavior evidenced by the formation of voids and fibrils was observed within the FPZ. On the other hand, quantitative predictions of fracture toughness were achieved for amorphous carbon. K-test specimen size was found to be critical in performing accurate measurements, i.e., the fully developed FPZ must be small relative to the MD domain size and the domain of the J-integral contours. Good agreement with available experimental data on Mode-I fracture toughness in amorphous carbon is observed, underscoring the ability of MD to predict fracture toughness in amorphous material from first principles. Owing to the intermediate fully developed FPZ length scale of amorphous carbon, it will serve as an important test-bed for studying nanoparticle reinforcement phenomena as well as fatigue and Mode-II loading in Chapters 9 and 10.

REFERENCES

1. Kausar, A., Role of thermosetting polymer in structural composite. *American Journal of Polymer Science and Technology*, 2017. **5**(1): p. 1–12.
2. Karnati, S.R., P. Agbo, and L. Zhang, Applications of silica nanoparticles in glass/carbon fiber-reinforced epoxy nanocomposite. *Composites Communications*, 2020. **17**: p. 32–41.
3. Lee, D.W. and B.R. Yoo, Advanced silica/polymer composites: Materials and applications. *Journal of Industrial and Engineering Chemistry*, 2016. **38**: p. 1–12.
4. Tsai, J.-L., H. Hsiao, and Y.-L. Cheng, Investigating mechanical behaviors of silica nanoparticle reinforced composites. *Journal of Composite Materials*, 2010. **44**(4): p. 505–524.
5. Poonpipat, Y., et al., Fracture behavior of silica nanoparticles reinforced rubber/epoxy composite. *Journal of Reinforced Plastics and Composites*, 2017. **36**(16): p. 1156–1167.

6. Hsieh, T., et al., The toughness of epoxy polymers and fibre composites modified with rubber microparticles and silica nanoparticles. *Journal of Materials Science*, 2010. **45**: p. 1193–1210.
7. Zhao, S., et al., Improvements and mechanisms of fracture and fatigue properties of well-dispersed alumina/epoxy nanocomposites. *Composites Science and Technology*, 2008. **68**(14): p. 2976–2982.
8. McGrath, L.M., et al., Investigation of the thermal, mechanical, and fracture properties of alumina–epoxy composites. *Polymer*, 2008. **49**(4): p. 999–1014.
9. Domun, N., et al., Improving the fracture toughness and the strength of epoxy using nanomaterials – A review of the current status. *Nanoscale*, 2015. **7**(23): p. 10294–10329.
10. Laurenzi, S., et al., Fracture mechanisms in epoxy composites reinforced with carbon nanotubes. *Procedia Engineering*, 2014. **88**: p. 157–164.
11. Shirodkar, N., S. Cheng, and G.D. Seidel, Enhancement of Mode I fracture toughness properties of epoxy reinforced with graphene nanoplatelets and carbon nanotubes. *Composites Part B: Engineering*, 2021. **224**: p. 109177.
12. Da Silva, D., et al., Application of amorphous carbon based materials as antireflective coatings on crystalline silicon solar cells. *Journal of Applied Physics*, 2011. **110**(4): p. 043510.
13. Kim, I.S., et al., Amorphous carbon films for electronic applications. *Advanced Materials*, 2023. **35**(43): p. 2204912.
14. Dodda, L.S., et al., LigParGen web server: An automatic OPLS-AA parameter generator for organic ligands. *Nucleic Acids Research*, 2017. **45**(W1): p. W331–W336.
15. Dodda, L.S., et al., 1.14* CM1A-LBCC: localized bond-charge corrected CM1A charges for condensed-phase simulations. *The Journal of Physical Chemistry B*, 2017. **121**(15): p. 3864–3870.
16. Jorgensen, W.L. and J. Tirado-Rives, Potential energy functions for atomic-level simulations of water and organic and biomolecular systems. *Proceedings of the National Academy of Sciences of the United States of America*, 2005. **102**(19): p. 6665–6670.
17. Odegard, G.M., et al., Predicting mechanical response of crosslinked epoxy using ReaxFF. *Chemical Physics Letters*, 2014. **591**: p. 175–178.
18. Li, C. and A. Strachan, Molecular dynamics predictions of thermal and mechanical properties of thermoset polymer EPON862/DETDA. *Polymer*, 2011. **52**(13): p. 2920–2928.
19. Jang, C.W., J.W. Mullinax, and J.W. Lawson, Mechanical properties and failure of aerospace-grade epoxy resins from reactive molecular dynamics simulations with nanoscale defects. *ACS Applied Polymer Materials*, 2022. **4**(8): p. 5269–5274.
20. Quinson, R., et al., Yield criteria for amorphous glassy polymers. *Journal of Materials Science*, 1997. **32**: p. 1371–1379.
21. Rottler, J. and M.O. Robbins, Yield conditions for deformation of amorphous polymer glasses. Physical Review E, 2001. **64**(5): p. 051801.
22. Bauwens, J.C., Yield condition and propagation of Lüders' lines in tension–torsion experiments on poly (vinyl chloride). *Journal of Polymer Science Part A-2: Polymer Physics*, 1970. **8**(6): p. 893–901.
23. Tack, J.L., *Thermodynamic and mechanical properties of EPON 862 with curing agent DETDA by molecular simulation*. 2010: Texas A & M University.
24. Zhou, Y., et al., Experimental study on the thermal and mechanical properties of multi-walled carbon nanotube-reinforced epoxy. *Materials Science and Engineering: A*, 2007. **452**: p. 657–664.
25. Tao, K., et al., Effects of carbon nanotube fillers on the curing processes of epoxy resin-based composites. *Journal of Applied Polymer Science*, 2006. **102**(6): p. 5248–5254.
26. Sun, L., et al., Mechanical properties of surface-functionalized SWCNT/epoxy composites. *Carbon*, 2008. **46**(2): p. 320–328.

27. Chevalier, J., et al., On a unique fracture micromechanism for highly cross-linked epoxy resins. *Journal of the Mechanics and Physics of Solids*, 2019. **122**: p. 502–519.
28. Khan, D., S. Singh, and A. Needleman, Finite deformation analysis of crack tip fields in plastically compressible hardening–softening–hardening solids. *Acta Mechanica Sinica*, 2017. **33**: p. 148–158.
29. Radue, M.S., et al., Comparing the mechanical response of di-, tri-, and tetra-functional resin epoxies with reactive molecular dynamics. *Journal of Polymer Science Part B: Polymer Physics*, 2018. **56**(3): p. 255–264.
30. Wang, C.H., *Introduction to fracture mechanics*. 1996: DSTO Aeronautical and Maritime Research Laboratory Melbourne, Australia.
31. Aditya, S.S. and S. Roy. Predicting fracture toughness in polymers using a novel atomistic-continuum concurrent coupling technique. in *ASME 2024 Aerospace Structures, Structural Dynamics, and Materials Conference*. 2024.
32. Khosrownejad, S.M., J.R. Kermode, and L. Pastewka, Quantitative prediction of the fracture toughness of amorphous carbon from atomic-scale simulations. *Physical Review Materials*, 2021. **5**(2): p. 023602.
33. Tersoff, J., Empirical interatomic potential for silicon with improved elastic properties. *Physical Review B*, 1988. **38**(14): p. 9902.
34. McCulloch, D., D. McKenzie, and C. Goringe, Ab initio simulations of the structure of amorphous carbon. *Physical Review B*, 2000. **61**(3): p. 2349.
35. Suk, J.W., et al., Mechanical measurements of ultra-thin amorphous carbon membranes using scanning atomic force microscopy. *Carbon*, 2012. **50**(6): p. 2220–2225.
36. Schultrich, B., et al., Elastic modulus of amorphous carbon films. *Physica Status Solidi (a)*, 1994. **145**(2): p. 385–392.
37. Ferrari, A., et al., Elastic constants of tetrahedral amorphous carbon films by surface Brillouin scattering. *Applied Physics Letters*, 1999. **75**(13): p. 1893–1895.
38. Jana, R., et al., Structural and elastic properties of amorphous carbon from simulated quenching at low rates. *Modelling and Simulation in Materials Science and Engineering*, 2019. **27**(8): p. 085009.
39. Li, X. and B. Bhushan, Evaluation of fracture toughness of ultra-thin amorphous carbon coatings deposited by different deposition techniques. *Thin Solid Films*, 1999. **355**: p. 330–336.
40. Li, X. and B. Bhushan, Measurement of fracture toughness of ultra-thin amorphous carbon films. *Thin Solid Films*, 1998. **315**(1–2): p. 214–221.
41. Fu, K., et al., Thickness-dependent fracture behaviour of amorphous carbon films on a PEEK substrate under nanoindentation. *Vacuum*, 2017. **144**: p. 107–115.

9 Crack-Tip Shielding Effect due to Embedded Nanoparticles in a Matrix Material

9.1 INTRODUCTION

In Chapter 8, the *K*-test and atomistic *J*-integral techniques were employed to quantitatively predict fracture toughness in amorphous carbon using MD simulations. One of the important applications of atomistic fracture simulation techniques is to enable rapid inverse design of nanocomposite materials with superior fracture properties. While epoxy nanocomposite materials have received a lot of attention and are important in various structural applications, the quantitative prediction of fracture toughness using MD simulations remains a challenge, as described in Chapter 8 due to FPZ size related length scale issues. Therefore, in this chapter, we present results from atomistic fracture simulations performed on amorphous carbon with embedded nanoparticles, since such simulations are feasible using currently available computing power due to the relatively modest size of the FPZ in amorphous carbon (~14 nm) as mentioned in Chapter 8. The results obtained for amorphous carbon-based nanocomposites are expected to qualitatively inform the design of epoxy nanocomposites as well. A variety of mechanisms have been proposed in the literature for nanoparticle-based fracture toughness enhancement which include crack pinning, crack bifurcation, and crack deflection [1–4]. However, as explained in Amirahmadi et al. [5], all of the above toughening mechanisms hinge on the fact that the crack is propagating or has already propagated. Therefore, they do not adequately explain enhancements in *initiation* fracture toughness prior to any crack propagation. A more convincing explanation of initiation toughness enhancement can be postulated using the concept of "crack-tip shielding," which is an intrinsic toughening mechanism, i.e., it occurs in the FPZ prior to crack growth. Crack-tip shielding due to microcrack formation ahead of a macroscale crack tip has been attributed as an important toughening mechanism in papers by Evans et al. [6] and Ravichandran et al. [7], and more recently by Liu et al. [8], using the multi-scale projection method. Essentially, when a nanocomposite material with a macroscale flaw is loaded in tension, it results in a region of stress concentration near the crack tip. Further, the presence of nanoparticles in the FPZ results in matrix/nanoparticle debonding due to tension, causing voids and micro-damage to form near the crack tip as depicted in Figure 9.1.

This local damage in the FPZ degrades the modulus of the material near the crack tip, which results in reduced stress intensity at the crack tip, thereby effectively

DOI: 10.1201/9781003440734-9

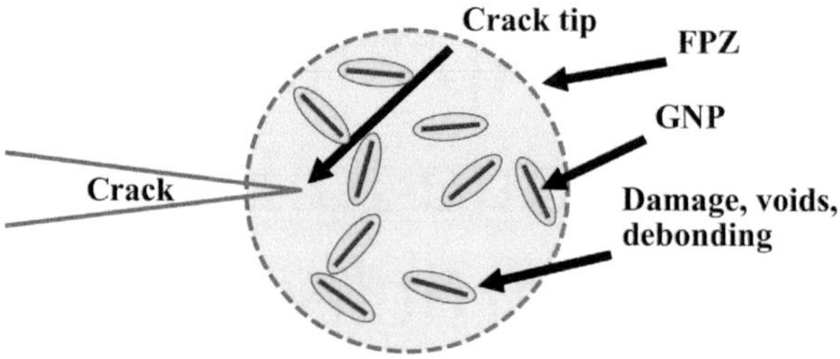

FIGURE 9.1 Schematic showing the formation of a local damage zone near the crack tip due to the debonding of the GNP from the matrix, resulting in crack-tip shielding effect.

delaying crack initiation. The relevant fracture mechanics equations justifying this mechanism are based on the fundamental concept of path-independence of the J-integral and are presented later in this chapter (refer to Equations (9.1a)–(9.1c)). While theoretical formulations exist for crack-tip shielding in the literature [6, 7, 9], to our knowledge, actual nanoscale validation has not been performed. In this chapter, our MD simulations provide definitive proof of crack-tip shielding effect in enhancing effective fracture toughness in nanocomposite materials, as well as the influence of nanoparticle orientation and location on shielding at crack initiation, and also during crack growth.

For epoxy nanocomposites, the addition of small weight percentage of graphene nanoplatelets (GNP) have resulted in significant fracture toughness enhancements (up to 227% enhancement in Mode-I fracture toughness K_{1c} and 600% improvement of fracture energy, G_{IC}) [5, 10]. Aside from particle weight percent, it was found that particle size played an important role in fracture toughness enhancement, with smaller particles exhibiting greater enhancement [5]. Particle orientation was also showed to be an important factor [2], something that can be controlled by electrical and magnetic fields [11]. Therefore, in this chapter, our focus is on the effect of GNPs on fracture toughness enhancement when embedded in an amorphous carbon matrix. A series of simulations were performed using different particle orientations and positions to investigate their effect on initiation fracture toughness and subsequent crack propagation.

9.2 MODELING METHODS AND SIMULATION PLAN

9.2.1 Methods for GNP-Embedded MD Simulations

The technique used for modeling amorphous carbon using MD simulations is identical to what is described in Chapter 8 (refer to Section 8.3.1). The Tersoff potential [12] used for modeling the amorphous carbon is also used to model the GNP and their interaction. The MD fracture simulations are conducted under plane stress conditions in the $X - Y$ plane, and allows Poisson's effect to occur in the out-of-plane direction (Z). A visualization of the $X - Y$ plane with GNPs embedded above and below a crack in the amorphous carbon matrix is shown in Figure 9.2.

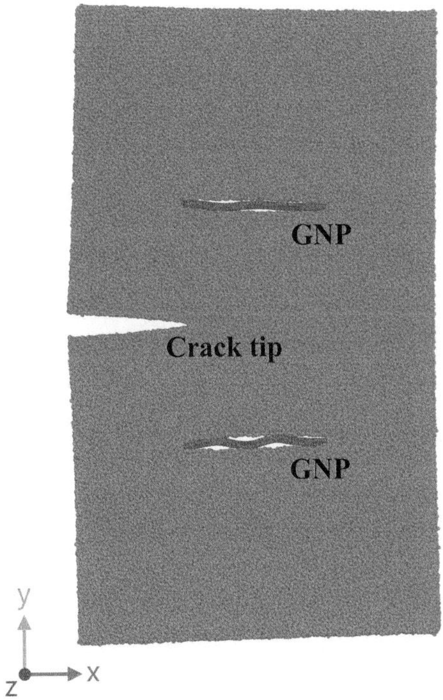

FIGURE 9.2 Visualization of GNPs embedded in amorphous carbon matrix material in the vicinity of a crack. Note debonding visible at the GNP/a-C interface at the top.

The *Z* direction is prescribed periodic boundary conditions, while the *X* and *Y* directions are nonperiodic. The GNPs are embedded at the start of the melt-quench procedure for amorphous carbon, but they are held "frozen" until the temperature is ramped down from 6000 to below 3150 K, which is lower than the glass transition temperature of amorphous carbon [13]. This is done so that the positions and orientations of the GNPs can be maintained as originally intended without energy equilibration-induced rotations, while at the same time allowing for local stress relaxation in the surrounding matrix. It should be noted that after the full annealing process, there is only nonbonded Van der Waals interaction between the GNPs and the carbon matrix. Chemical functionalization of the GNP was not considered here, because weak interfacial bonding actually facilitates crack-tip shielding via particle debonding. This is a departure from the more traditional approach of making the GNP interface stronger using covalent bonds, which improves nanocomposite stiffness, but not necessarily its toughness.

9.2.2 Fracture Simulation Test Matrix for GNP Embedded in Amorphous Carbon

The *K*-test geometry is maintained constant in all the MD fracture simulation cases, with dimensions equal to the "medium-sized" specimen used in Chapter 8, i.e., 24 × 40 nm MD domain with an 8 nm crack in amorphous carbon (a-C). All GNPs are of the same

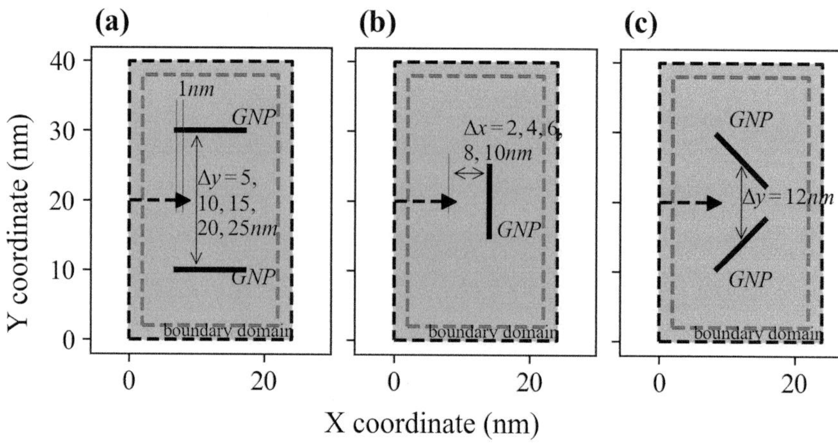

FIGURE 9.3 Schematic of fracture simulation test configurations in a-C with horizontal, vertical, and 45-degree oriented GNP cases shown in (a), (b), and (c), respectively. Black arrow indicates crack tip location. Also shown are various GNP separation distances that were simulated.

size in this study with an in-plane length of 10 nm and out-of-plane width of 2.4 nm (equal to the simulation-box depth in the Z direction). The thickness of each GNP particle is approximately 0.34 nm. Three orientations of the GNPs are studied—horizontal, vertical, and 45° – as shown in Figure 9.3.

For the horizontal and vertical orientations, five positions with respect to the crack tip are studied, therefore comprising a total of eleven simulation cases, including the baseline case without GNPs. For each test case, three independent MD realizations of the specimen by the melt-quench procedure are generated, followed by fracture simulations. This is done to capture statistical variation in the results from stochastic variations in the amorphous carbon morphology and MD simulations at elevated temperature, analogous to multiple real-life fracture experiments being performed to obtain statistically meaningful failure data. In the present study, only rotation of the GNP in the $X - Y$ plane is considered, as shown in Figure 9.3. Out-of-plane rotated orientations of the GNPs are not considered in order to keep the number of simulations tractable. All MD simulations in this chapter are performed at $T = 300$ K.

9.2.3 Predicting Fracture Toughness Using K-Test and J-Integral

The general K-test procedure is the same as that used in Chapter 8, with K-field displacement loading applied to a 2 nm boundary domain using SMD. Because the K-test assumes the simulation domain is only a subset of a much larger elastic domain, we must factor in the effect of GNPs dispersed in the elastic domain on the far-field elastic modulus. Any potential changes to the elastic modulus will affect the K-field displacements, which are calculated based on the equations described in Chapter 5. To estimate the effect of GNPs on the elastic modulus of the amorphous carbon nanocomposite, a modified Halpin-Tsai analysis was performed as presented in [14]. Halpin-Tsai analysis is more accurate for computing modulus enhancement of a matrix material due to embedded nanoparticles than the traditional rule-of-mixtures. Isotropic material behavior was assumed in the carbon material away from the crack tip.

Based on the present weight percent of GNPs used (~0.5%), the calculated increase in elastic modulus based on the Halpin-Tsai analysis is ~0.3%. Therefore, in this specific situation, the elastic modulus is not significantly impacted by the presence of GNPs, and therefore it does not need to be factored into the K-field displacement calculations. However, for other combinations of matrix material, nanoparticle, and weight percent, there might be a considerable impact on the nanocomposite's elastic modulus, and consequently, the K-field boundary displacements will need to be calculated using revised nanocomposite material properties rather than those of pure a-C. Nanoparticle-induced anisotropy is also a possibility and may need to be incorporated into the K-test boundary condition in the case of a large number of nanoparticles embedded in the matrix, but it is ignored here because only a few embedded GNPs are considered.

The atomistic J-integral was used to evaluate the crack initiation and propagation in the GNP-based a-C nanocomposite simulations. The J-integral contour was placed as close to the boundary domain as feasible without having computation boxes overlapping with the boundary domain. Moreover, the GNPs in all cases were fully contained within the J-integral contour, so that any inelastic damage originating due to the debonding of the GNPs is captured by the J-integral.

9.3 EFFECT OF GNPS ON INITIATION FRACTURE TOUGHNESS

9.3.1 Nanoparticle Orientation Effect

Three GNP orientations were considered—horizontal, vertical, and 45°, as depicted in Figure 9.3. The three cases considered changed the stress distribution around the crack tip and the J-integral in profoundly different ways, leading to dramatic differences in initiation fracture toughness. The von Mises stress contour indicating the MD-computed FPZ for the baseline case is shown in Figure 9.4a without any GNPs.

The contour of the fully developed FPZ when two GNPs are placed parallel to the crack tip is shown in Figure 9.4b with a separation distance of 10 nm between the GNPs. Figure 9.4c shows the contour of the fully developed FPZ for the case when the horizontal GNPs are rotated and placed at an angle to 45° to the crack tip. Finally, Figure 9.4d shows the contour of the fully developed FPZ when a single GNP is placed vertically ahead of the crack tip at a distance of 4 nm. The qualitative effect of the GNP orientation on the von Mises stress field at the crack tip is visualized in Figure 9.4a–d. All of the von-Mises stress contours are presented at a Mode-I far-field stress intensity of $K_I = 5.5\ \mathrm{MPa}\sqrt{\mathrm{m}}$, which was predicted in Chapter 8 to be close to the baseline fracture toughness for amorphous carbon using the present methods. Stress contour lines are drawn at the yield strength of amorphous carbon, which was found to be 22 GPa in Chapter 8, thereby delineating the boundary of the FPZ. Both MD and LEFM predicted FPZs are shown for comparison for baseline

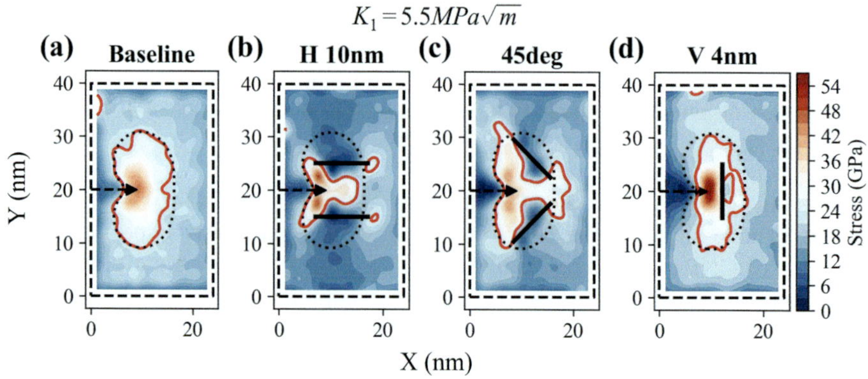

FIGURE 9.4 von-Mises stress contour around the crack tip for baseline and various orientations of GNPs at $K_{\rm I} = 5.5$ MPa$\sqrt{\rm m}$. Red contour lines are FPZ from MD, and black dotted contour lines are FPZ contour predicted by linear elastic fracture mechanics (LEFM).

and GNP cases. Red contour lines are from MD simulation for each case and the black dotted contour lines are from LEFM prediction in Figure 9.4. It is apparent that the size and shape of the FPZ for the three GNP orientations shown in Figure 9.4b–d are very different compared with each other and with the baseline case. Evidently, the horizontal GNPs have a shielding effect on the crack as they decrease the von-Mises stress contour near the crack tip, thereby significantly reducing the size of the FPZ at $K_{\rm I} = 5.5$ MPa$\sqrt{\rm m}$. On the other hand, the vertical GNP appears to intensify the stress concentration at the crack tip (shown as dark red in Figure 9.4d), with a reduction in stress immediately to the right of the GNP. As one might expect, the 45-degree case offers an intermediate stress state between the horizontal and vertical cases. A quantitative analysis of stress distribution is presented in Figure 9.5 by plotting von-Mises, volumetric, and normal-Y stress components ahead of the crack tip for all three GNP cases and at $K_{\rm I} = 5.5$ MPa$\sqrt{\rm m}$.

The LEFM data are shown by the dotted line for comparison. The gray shaded area in Figure 9.5 is the projection of the span of the 10 nm horizontal GNP on the crack plane. Similarly, the vertical black line indicates the location of the vertically oriented GNP relative to the crack tip. As can be seen, the horizontal GNP case offers significant shielding by lowering the von-Mises, volumetric, and normal-Y stress peaks by 22%, 17%, and 25%, respectively relative to the baseline. On the other hand, the vertical GNP results in a "negative-shielding" effect with von-Mises stress 18% above the baseline case. Hence, the key takeaway here is that horizontal and vertical orientations of GNPs near the crack tip play a pivotal role in local stress redistribution. However, the true measure of crack-tip shielding effects due to GNPs comes from actual visualization of crack initiation as shown in Figure 9.6, where $H10$ implies horizontal GNPs with 10 nm separation, and $V4$ implies a vertical GNP at a distance of 4 nm from the crack tip.

The results are quite interesting. While the baseline case has clearly initiated at $K_{\rm I} = 6$ MPa$\sqrt{\rm m}$ with significant crack propagation at higher loading levels, crack

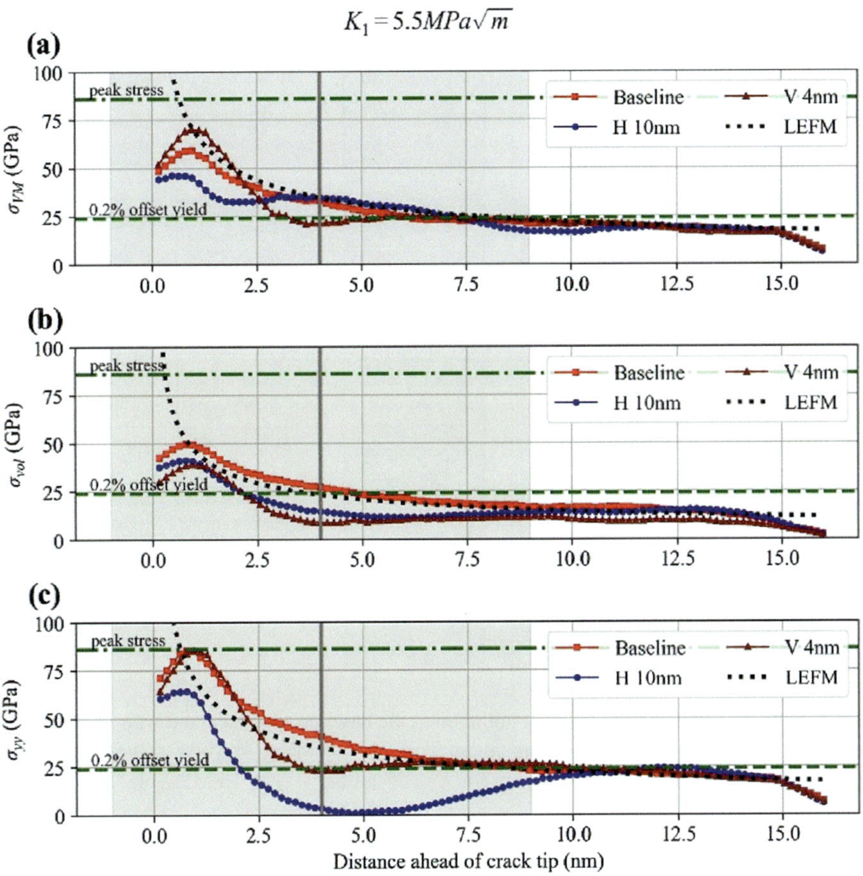

FIGURE 9.5 Stress components (von-Mises (top), volumetric (middle), and normal-Y (bottom)) plotted as a function of distance from crack tip for baseline and various GNP orientations.

initiation in the presence of horizontal and 45-degree GNPs is not observed. Instead of fracture at the main crack, debonding of GNPs from the matrix is observed for the horizontal and 45° cases, which is the main mechanism of crack-tip shielding, resulting in local dissipation of strain energy. On the other hand, crack initiation in the vertical GNP case occurs at a lower loading level similar to the baseline case. After the crack tip reaches the vertical GNP, however, further crack growth is retarded by the vertical GNP by a crack-deflection/pinning mechanism as can be seen in Figure 9.6. It is important to note that crack initiation in each simulation is quantitatively identified by identifying the load level (K_{I}) at which the maximum principal stress at the crack tip reaches a peak value and then begins to decline. Based on this, the critical stress intensity factor (K_{1c}) for each simulation case was identified and it is shown in Figure 9.7 for different cases of GNP orientation and separation distance, with H implying horizontal and V vertical orientation, and the

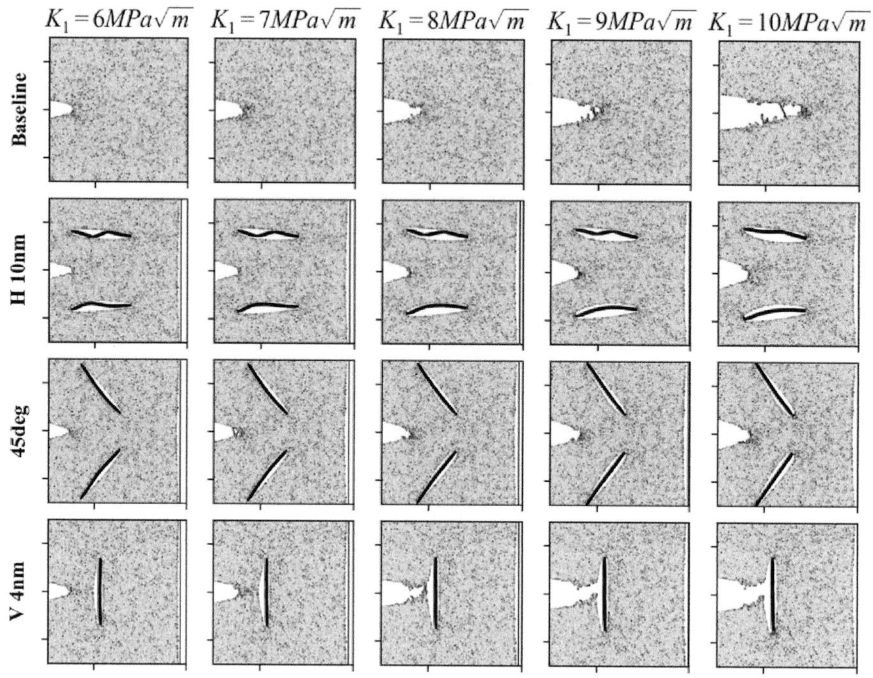

FIGURE 9.6 Visualization of crack initiation in the baseline specimen and with various GNP orientations. Crack initiation occurs at applied $K_I = 6\ \mathrm{MPa}\sqrt{\mathrm{m}}$ (first column), and the remaining columns depict crack growth for $K_I = 7$ all the way to $10\ \mathrm{MPa}\sqrt{\mathrm{m}}$.

indexing number implying separation distance for H, and distance from the crack tip for V.

A summary of the mean K_{Ic} data presented in Figure 9.7 can be found in Table 9.1. The error bars signify the standard deviation from the mean from three MD realizations for each case. Corroborating our observations from the near-tip stress profile in Figure 9.5 and crack-tip visualization in Figure 9.6, all of the horizontal GNP cases in Figure 9.7 result in enhanced fracture toughness of the nanocomposite relative to the baseline case (red line), whereas almost all of the vertical GNP cases degrade the initiation fracture toughness. The 45-degree case has a mild enhancing effect on initiation toughness and falls between the vertical and horizontal cases, as one might expect. In this study, the GNPs are located ahead of the crack-tip for all MD simulations. Recent work by Liu et al. [8] indicates that horizontal microcracks or flaws located behind the crack tip and parallel to the crack plane can have an even greater shielding effect on the macrocrack than those in front of it. To gain further insight into the GNP-controlled failure mechanism, the atomistic J-integral evaluated for horizontal and vertical GNP orientations is plotted against applied K_I loading in Figure 9.8 and compared with the baseline (no GNP) case. The horizontal GNP case ($H10$, shown as the blue curve) deviates early from the LEFM prediction (dashed line) at around $K_I = 2\ \mathrm{MPa}\sqrt{\mathrm{m}}$ indicating the onset of GNP debonding, and

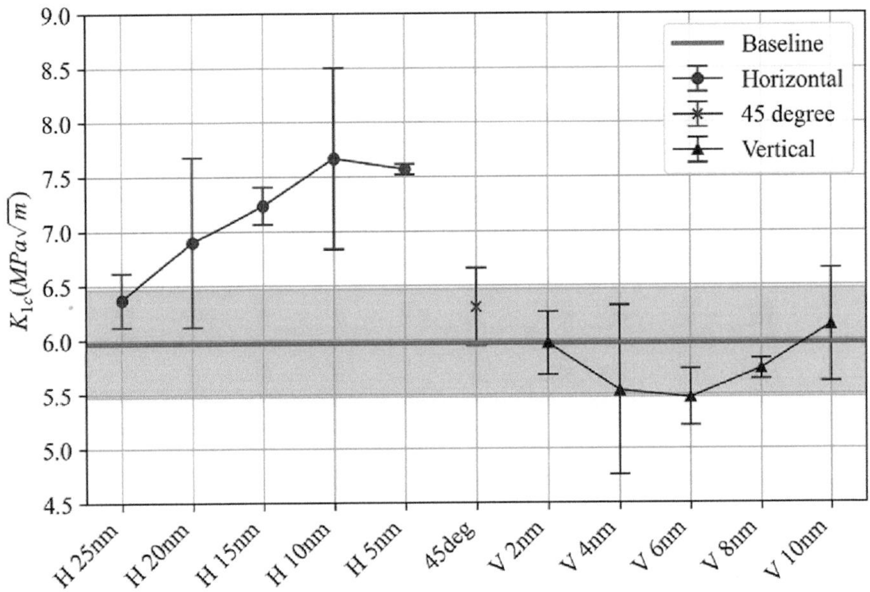

FIGURE 9.7 Summary of MD-predicted critical stress intensity factor (K_{1c}) in a-C for all simulation cases including baseline and with various GNP orientations and separation distances.

TABLE 9.1
Summary of Fracture Toughness and Near-Tip Moduli for All Simulation Cases with Various GNP Orientations and Position/Separation

GNP Orientation Case	$K_{1c}(\text{MPa}\sqrt{\text{m}})$	$K_{1f}(\text{MPa}\sqrt{\text{m}})$	$E_{\text{ntip}}(\text{GPa})$	$K_{1c,\text{ntip}}(\text{MPa}\sqrt{\text{m}})$
Baseline (no GNP)	6.0 ± 0.5	10.0 ± 1.4	461 ± 17	6.0 ± 0.5
*H*25 nm	6.4 ± 0.2	12.7 ± 0.9	279 ± 5	5.0 ± 0.2
*H*20 nm	6.9 ± 0.8	14.8 ± 1.0	270 ± 10	5.4 ± 0.7
*H*15 nm	7.2 ± 0.2	14.0 ± 0.0	267 ± 13	5.6 ± 0.3
*H*10 nm	7.7 ± 0.8	13.7 ± 1.2	257 ± 15	5.8 ± 0.5
*H*5 nm	7.6 ± 0.0	12.7 ± 0.9	305 ± 30	6.2 ± 0.3
*V*2 nm	6.0 ± 0.3	12.8 ± 1.2	465 ± 32	6.1 ± 0.3
*V*4 nm	5.5 ± 0.8	13.8 ± 1.3	428 ± 30	5.4 ± 0.8
*V*6 nm	5.5 ± 0.3	14.2 ± 0.6	470 ± 17	5.6 ± 0.2
*V*8 nm	5.7 ± 0.1	16.3 ± 2.1	473 ± 21	5.9 ± 0.2
*V*10 nm	6.1 ± 0.5	15.3 ± 0.5	476 ± 19	6.3 ± 0.4
45°	6.3 ± 0.4	12.3 ± 0.6	306 ± 18	5.2 ± 0.2
Experimentally measured range	2.4–6.0	—	100–700	—

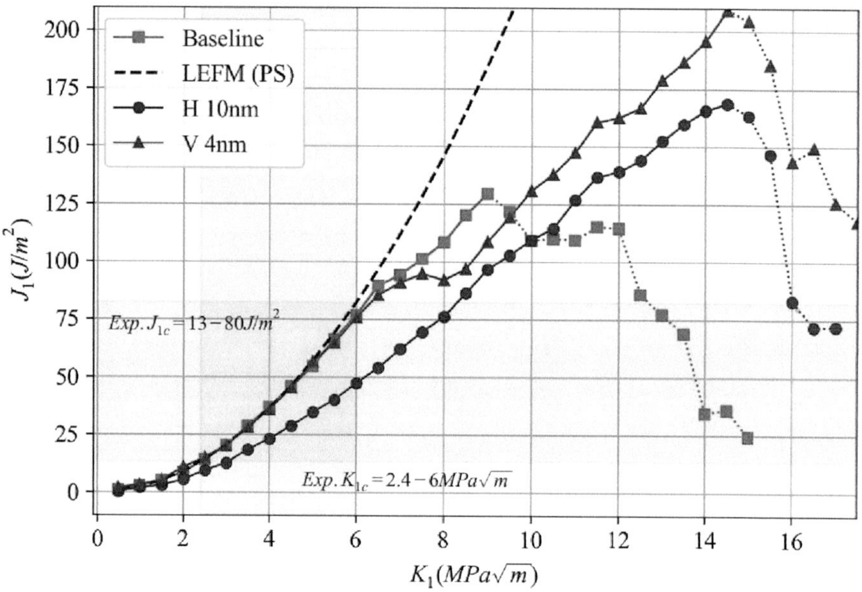

FIGURE 9.8 MD-predicted J-integral vs. K_1 for baseline and various GNP orientations at $T = 300$ K

exhibits J-integral values significantly lower than the vertical GNP, the baseline case, and the LEFM prediction, respectively. This agrees with the fact that crack-tip shielding effects originate from strain-energy dissipation due to nanoparticle/matrix debonding for the horizontal GNP case. Because of damage related energy dissipation due to debonding within the J-integral domain, there is less strain energy available to be released, i.e., lower SERR which can be interpreted as crack-tip shielding, which is reflected in Figure 9.8.

Of course, the LEFM curve does not include any effects of GNP debonding and hence shows no decline. The J-integral value for the single vertically oriented GNP (burgundy curve in Figure 9.8) on the other hand follows the baseline and the LEFM curves up until $K_1 = 5.4$ MPa$\sqrt{\text{m}}$, when deviation occurs due to energy dissipation caused by crack initiation. The good agreement with LEFM for the vertical GNP case underscores the absence of significant crack shielding, and in fact, negative shielding is observed due to stress concentration caused by the vertical GNP 4 nm from the crack tip. This observation runs contrary to the popular misconception that vertically oriented GNPs are good for initiation toughness enhancement in materials. The J-integral curve for all three cases continues to increase after crack initiation, decreasing only when the crack grows beyond the J-integral contour, near the edge of the simulation domain. This can be viewed as a point of complete failure of the structure and gives some insight into the crack propagation rate. While not a standard fracture metric, K_{1f} can be used to denote this loading level when complete failure has occurred, corresponding to the peak of the J-integral curve in Figure 9.8.

9.3.2 Analytical Model for Crack Tip Shielding Effect

A summary of the SMD results from all the simulation cases, for various orientations and positions of GNPs, is shown in Table 9.1. The columns of the table are the critical stress intensity factor (K_{1c}), the loading level at which the crack propagates entirely through the specimen (K_{1f}), the near crack tip modulus (E_{nc}) which includes the effect of debonding, and the near crack-tip crack stress intensity factor $(K_{1c,\text{nc}})$. The near crack-tip modulus, E_{nc}, is approximately measured based on local stress and strain computed using MD along the crack plane and gives a quantitative measure of the stiffness degradation caused by the local damage induced by the GNP/matrix debonding and other nonlinear, inelastic phenomena. As noted previously, a degradation of local elastic modulus also reduces the stress intensity and SERR viewed locally by the crack tip, leading to crack-tip shielding as discussed in the next paragraph.

The mathematical formalism for crack-tip shielding is as follows, and more details can be found in Evans et al. [6]. For a linear elastic isotropic material under plane stress, the relationship between the J-integral and the stress intensity factor (K) is given by,

$$J = \frac{K_1^2}{E} \tag{9.1a}$$

where E is the Young's modulus of the material. As described by Rice [15] and Eshelby [16], the path-independence of the J-integral within the LEFM framework leads to, for the plane stress case,

$$J_{1c,\text{ntip}} = \frac{K_{1c,\text{ntip}}^2}{E_{\text{ntip}}} = J_{1c,\text{ff}} = \frac{K_{1c,\text{ff}}^2}{E_{\text{ff}}} \tag{9.1b}$$

$$\Rightarrow K_{1c,\text{ff}} = K_{1c,\text{ntip}}\sqrt{\frac{E_{\text{ff}}}{E_{\text{ntip}}}} \tag{9.1c}$$

where $J_{1c,\text{ntip}}$ and $J_{1c,\text{ff}}$ are the near crack tip and far-field J_{1c}, respectively, and E_{ntip} and E_{ff} are the near crack tip and far-field elastic moduli of the matrix material respectively. Now, in Equation (9.1c), if E_{ntip} is lower than E_{ff} due to GNP-debonding related damage, the far-field stress intensity factor $K_{1c,\text{ff}}$ becomes greater than the near-tip $K_{1c,\text{ntip}}$. Therefore, a higher far-field fracture toughness is measured, even though the local fracture toughness near the crack tip $(K_{1c,\text{ntip}})$ has not changed. Hence, the ratio $\sqrt{\frac{E_{\text{ff}}}{E_{\text{ntip}}}}$ quantifies the shielding effect. In Table 9.1, we find that E_{nc} is indeed degraded relative to the baseline for most of the horizontal GNP cases, whereas E_{nc} is slightly greater than or equal to the baseline for most of the vertical GNP cases, therefore resulting in "negative-shielding" effect. Based on E_{nc} and K_{1c} values reported in Table 9.1, the near crack tip toughness $K_{1c,\text{ntip}}$ was calculated as a cross-check and values were found to be reasonably close to the baseline value of 6 MPa$\sqrt{\text{m}}$, as expected. It should be noted that Equation (9.1a) can easily be modified to include plane strain conditions.

9.3.3 Effect of GNP Separation Distance

From the MD simulation results presented in the previous section, the effect of GNP orientation was clearly demonstrated – horizontal GNPs provide significantly better initiation fracture toughness enhancement relative to vertical GNPs. In this section, we take a more detailed look at how the separation distance between the horizontal GNPs affects the stress distribution and initiation fracture toughness. Von-Mises stress contour plots for three of the five MD simulation cases with horizontal GNPs, namely $H5, H15,$ and $H25$, are shown in Figure 9.9, at a far-field loading of $K_{\rm I} = 5.5\ \text{MPa}\sqrt{\text{m}}$. The dashed lines in Figure 9.9 indicate the fully developed FPZ contour predicted using LEFM.

As is evident by the red contour plots in Figure 9.9, the FPZ is severely constrained between the lower and upper GNPs. von-Mises stress bands form, stretching from the crack tip to the left-hand tips of the GNPs in all three cases, and there is a significant reduction in the overall process zone area due to crack-tip shielding. In the 15 nm separation case, an "X-shaped" FPZ contour is observed from MD simulation. Figure 9.10a–c show the von-Mises, volumetric, and normal-Y stress components, respectively, as a function of distance from the crack tip for horizontal GNPs at various separation distances at far-field loading of $K_{\rm I} = 5.5\ \text{MPa}\sqrt{\text{m}}$.

Again, the gray shaded area in Figure 9.10 is the projection of the span of the 10 nm horizontal GNP on the crack plane. Based on the quantitative stress plots ahead of the crack tip shown in Figure 9.10, it is clear that all the horizontal GNP cases are effective in shielding the crack tip by reducing stress levels relative to the baseline case. The 5 nm GNP separation resulted in the most significant reduction in stress, down by almost 50%. The 15 nm separation was the next best case, followed by the 25 nm case. This is expected, because the further the horizontal GNPs are from the crack, the less the shielding effect is and the closer the stress pattern is to

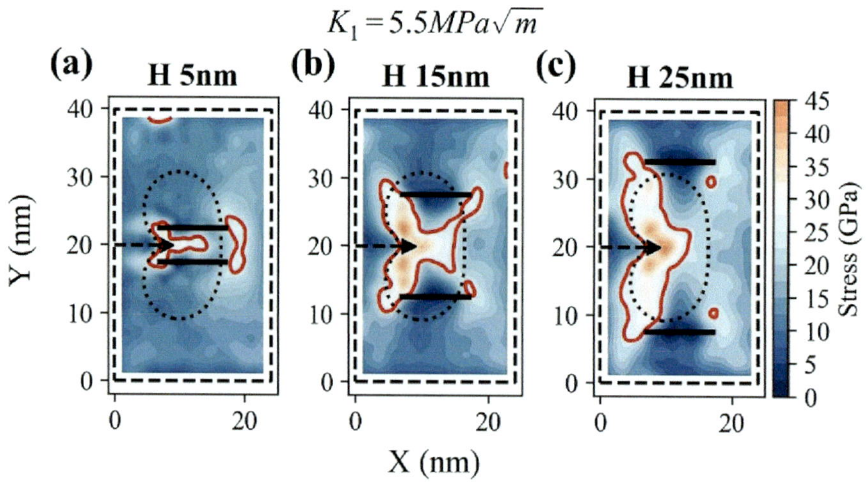

FIGURE 9.9 von-Mises stress contours near the crack tip for horizontal GNPs with varying separation distance at $K_{\rm I} = 5.5\ \text{MPa}\sqrt{\text{m}}$.

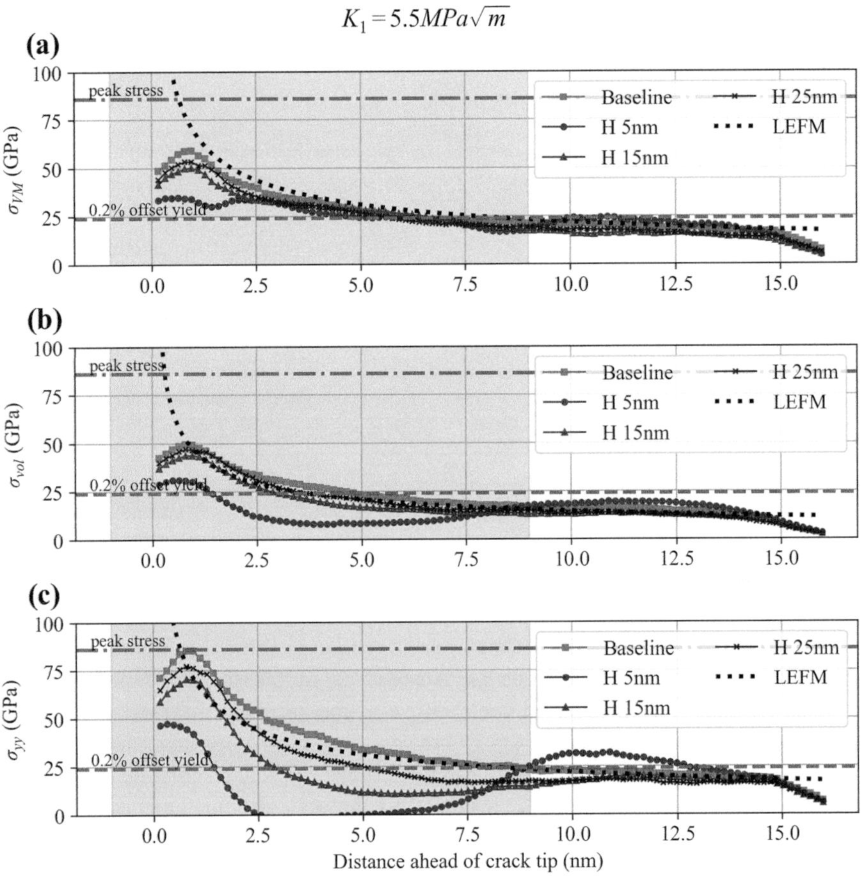

FIGURE 9.10 von-Mises, volumetric, and normal-Y stress components ahead of the crack tip for horizontal GNPs at various separation distances at $K_1 = 5.5$ MPa$\sqrt{\text{m}}$.

the baseline case. It is interesting to note that the normal-Y stress component in Figure 9.10c goes slightly negative to equilibrate the sharp rise near the crack tip.

9.3.4 Influence of GNP Position and Separation Distance on Crack Propagation

As demonstrated in Section 9.3.1 for crack initiation, visualization of crack propagation provides more concrete evidence of crack-tip shielding and toughness enhancement, as shown in Figure 9.11 for horizontal GNPs with various separation distances. Several interesting observations can be made. As described in Section 9.3.1, crack initiation in the 5 and 15 nm cases is delayed compared with the 25 nm case, indicating better shielding and toughening as is expected based on stress plots. However, after crack initiation, there is significant crack deflection and bifurcation that occur in the 5 and 15 nm cases, as can be seen in Figure 9.11.

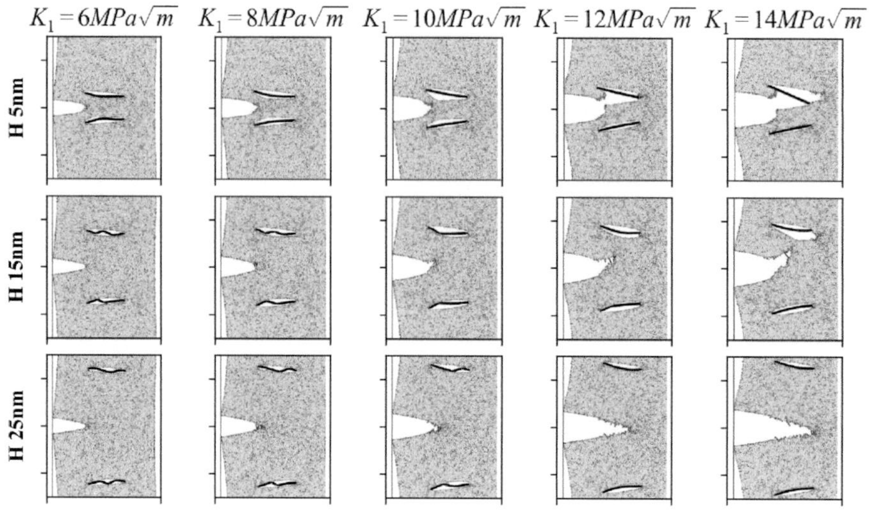

FIGURE 9.11 Visualization of crack initiation and propagation for horizontal GNPs with various separation distances at different applied K_I values.

While only crack deflection in the upper direction is shown in Figure 9.11, both upper- and lower-side deflections were observed in the three independent MD realizations that were performed for each case, underscoring the stochastic nature of the crack-growth process. For the 25 nm case, the crack propagates almost straight ahead as one might expect under pure Mode-I loading, closely resembling the baseline case. From the data in Figure 9.7, it is evident that an optimum separation distance exists in terms of maximizing far-field K_{1c}, due to shielding, which in this case is the horizontal GNPs with a separation of 10 nm. At lower separations, it is observed that the stress concentration at the debonding surrounding the GNPs begins to interact with the crack tip, thereby acting as an attractor and effectively lowering the local matrix toughness. Further investigation is underway to confirm these observations.

Based on data in Figure 9.7, from a crack-initiation point of view, the vertical GNPs were found to underperform relative to the horizontal GNPs. However, it should be noted that vertical GNPs have the potential to act as a barrier to crack propagation via crack pinning and crack deflection. Therefore, the effect of the distance of a vertical GNP from the crack tip on crack propagation behavior is visualized in Figure 9.12, for a vertical GNP with 2 nm distance from the crack tip (row 1), for a vertical GNP with 6 nm distance from the crack tip (row 2), and for a vertical GNP with 10 nm distance from the crack tip (row 3), respectively.

After crack initiation at $K_I = 6$ MPa$\sqrt{\text{m}}$ given by column 1 in Figure 9.12, as the applied K value is increased, the crack begins to propagate and impinges on each of the vertical GNPs regardless of distance from the crack tip, as shown in the subsequent columns in Figure 9.12. Therefore, the vertical GNP with 2 nm distance from the crack tip does a good job of arresting the crack from growing further, until a much higher loading level when the crack bifurcates and grows around it. Similar crack-pinning behavior is observed for the vertical GNP located at 6 and 10 nm from the crack tip, respectively, in Figure 9.12.

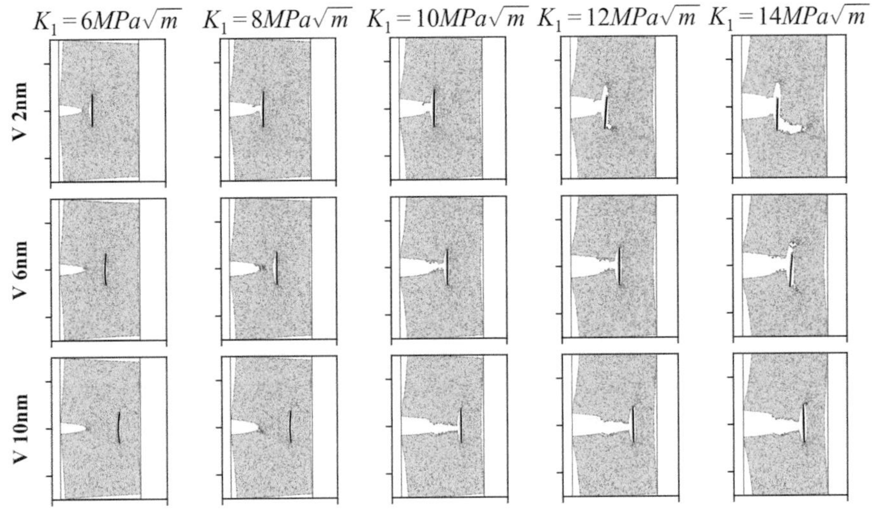

FIGURE 9.12 Visualization of crack initiation and propagation for vertical GNPs with various separation distances at different applied K_I values.

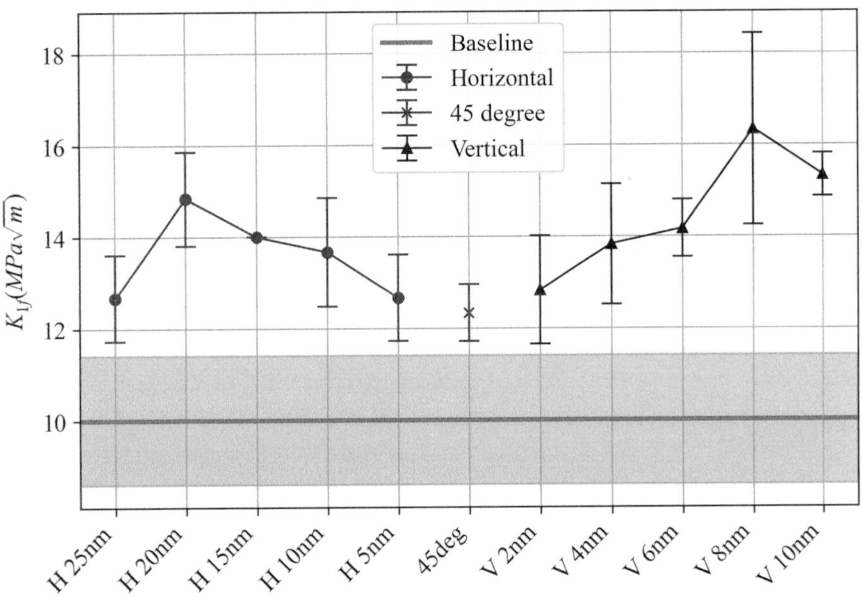

FIGURE 9.13 MD-computed stress intensity factor at complete fracture (K_{1f}) of a-C for all simulation case including baseline case and with various GNP orientations and positions.

While the main interest of this study was GNP-based enhancement of initiation fracture toughness, i.e., K_{1c}, it is definitely of interest to understand how crack propagation is affected by the various GNP cases. In Figure 9.13, the K_{1f} metric, intended to quantify the K_I value at full fracture of the MD specimen, is shown for all GNP orientations and positions and compared with the baseline case.

As anticipated, the vertical GNPs performed better than or equal to the horizontal GNPs in term of crack propagation resistance, even though their effect on crack initiation toughness is significantly worse. Therefore, in the inverse design of nanocomposites, it is important to consider the tradeoff between crack initiation toughness and crack propagation resistance. Our preliminary results here indicate that a random distribution of horizontally and vertically oriented GNPs in the matrix material is likely to provide the optimum overall toughness enhancement.

9.4 SUMMARY

The objective of this chapter was to apply the quantitative fracture toughness prediction tools developed so far to the design of nanostructured composite materials. In this case, amorphous carbon (a-C) was used as a suitable test bed, exhibiting measurable inelastic behavior near the crack tip while still being computational tractable due to a smaller FPZ size of ~12 nm. GNPs, which were used with epoxy resin to dramatically enhance fracture toughness in the literature, were used here with a-C to study the effect of GNP on crack-tip shielding. The essential crack initiation toughening mechanism proposed in the literature—crack-tip shielding—was carefully studied using stress plots, K_1, J-integral, and crack-growth visualization. Quantitative insights regarding GNP orientation and position/separation distance on initiation fracture toughness and crack propagation were presented. Horizontal GNPs were found to improve fracture toughness K_{1c} by up to ~25% compared with the baseline case (without GNPs). On the other hand, vertical GNPs exhibited "negative-shielding" effects, lowering the K_{1c} below baseline. Both horizontal and vertical GNPs, however, exhibited interesting crack-deflection and arresting mechanisms to slow down crack propagation relative to the baseline case once the crack had initiated. We believe the present study sets the stage for a wide range of MD-based explorations of the design space utilizing nanoparticle-based tailoring of toughening mechanisms in materials, with special focus on nanoparticles positioned behind the main crack tip.

REFERENCES

1. Chandrasekaran, S., et al., Fracture toughness and failure mechanism of graphene based epoxy composites. *Composites Science and Technology*, 2014. **97**: p. 90–99.
2. Quaresimin, M., et al., Toughening mechanisms in polymer nanocomposites: From experiments to modelling. *Composites Science and Technology*, 2016. **123**: p. 187–204.
3. Park, Y.T., et al., Epoxy toughening with low graphene loading. *Advanced Functional Materials*, 2015. **25**(4): p. 575–585.
4. Rafiee, M.A., et al., Fracture and fatigue in graphene nanocomposites. *Small*, 2010. **6**(2): p. 179.
5. Amirahmadi, S., S. Aditya, and S. Roy, An experimental-analytical investigation of particle size effect on the mode I fracture behavior of polymer/graphene nanocomposites. *Composite Structures*, 2024. **347**: p. 118420.
6. Evans, A., S. Williams, and P. Beaumont, On the toughness of particulate filled polymers. *Journal of Materials Science*, 1985. **20**: p. 3668–3674.

7. Ravichandran, G. and C. Liu, Crack tip shielding in elastic particulate composites undergoing damage. *Engineering Fracture Mechanics*, 1998. **59**(6): p. 713–723.
8. Liu, G., Dai Zhou, Yan Bao, Jin Maa, Zhaolong Han, Multiscale analysis of interaction between macro crack and microdefects by using multiscale projection method. *Theoretical and Applied Fracture Mechanics*, 2017. **90**: p. 65–74.
9. Ortiz, M., Microcrack coalescence and macroscopic crack growth initiation in brittle solids. *International Journal of Solids and Structures*, 1988. **24**(3): p. 231–250.
10. Roy, S. and A. Kumar, Effect of particle size on mixed-mode fracture of nanographene reinforced epoxy and mode I delamination of its carbon fiber composite. *Composite Structures*, 2017. **181**: p. 1–8.
11. Oliva-Avilés, A., F. Avilés, and V. Sosa, Electrical and piezoresistive properties of multi-walled carbon nanotube/polymer composite films aligned by an electric field. *Carbon*, 2011. **49**(9): p. 2989–2997.
12. Tersoff, J., Empirical interatomic potential for silicon with improved elastic properties. *Physical Review B*, 1988. **38**(14): p. 9902.
13. Best, S., et al., Evidence for glass behavior in amorphous carbon. *C*, 2020. **6**(3): p. 50.
14. Shokrieh, M.M., et al., Stiffness prediction of graphene nanoplatelet/epoxy nanocomposites by a combined molecular dynamics–micromechanics method. *Computational Materials Science*, 2014. **92**: p. 444–450.
15. Rice, J.R., A path independent integral and the approximate analysis of strain concentration by notches and cracks. *Journal of Applied Mechanics*, 1968. **35**: p. 379–386.
16. Eshelby, J., The elastic energy-momentum tensor. *Journal of Elasticity*, 1975. **5**: p. 321–335.

10 Mode-II and Mode-I Fatigue Loading in Amorphous Carbon Using Steered-MD Based *K*-Test Approach

10.1 INTRODUCTION

In the literature on atomistic fracture modeling, studies have primarily focused on Mode-I loading. Only a limited number of studies have been conducted on fracture modeling for Mode-II loading, which necessitates pure shear loading conditions at the boundaries or a *K*-test approach as described in Chapter 5. *K*-test-based Mode-II and mixed-mode fracture simulations were performed by researchers on material systems such as crystalline silicon [1], silicon-silicon interface [2], and monolayer boron nitride [3]. Other Mode-II fracture simulations were performed using far-field shear boundary conditions on idealized brittle materials and interfaces [4]. To our knowledge, no studies on Mode-II fracture simulations on amorphous carbon exist in the literature. Therefore, one of the objectives of this chapter is to apply the SMD-based *K*-test methods presented in previous chapters to modeling and predicting Mode-II fracture in amorphous carbon at room temperature ($T = 300$ K). Primarily, a study of the MD domain size effect on Mode-II crack initiation and propagation toughness prediction is presented in this chapter.

A second objective of this chapter is to apply the *K*-test techniques to simulate Mode-I fatigue crack growth behavior in amorphous carbon. Fatigue crack growth has been studied by researchers using MD simulations for various crystalline materials including magnesium [5], iron [6], nickel [7], and titanium [8]. In all the referenced studies on fatigue crack propagation in crystalline materials, a center-cracked tension test was used with cycle counts in the order of 10 cycles. While crack propagation was observed in all the referenced studies, a definitive analysis of the FPZ size relative to the MD specimen domain size at the applied peak loads was not presented, and therefore, the validity of these studies from a *K*-test perspective was not clearly established. For amorphous materials, studies on fatigue loading in uniaxial tension/compression have been performed by researchers based on RVE for uncracked polymers and polymer composites [9, 10]. However, to our knowledge, no studies have been conducted on fatigue in cracked specimens in amorphous materials using MD simulations. Moreover, no such studies have been attempted

DOI: 10.1201/9781003440734-10

for amorphous carbon. Therefore, in this chapter, Mode-I fatigue crack growth simulations are conducted in conformance with the K-test boundary conditions, and at room temperature ($T = 300$ K).

10.2 METHODS

The K-test methodology, geometry of the specimens, and synthesis procedure for amorphous carbon (a-C) used in this chapter are similar to the methods presented in Chapter 8 for the Mode-I case. For Mode-II loading, the K-test displacement fields (u_x and u_y) applied to the MD domain boundaries are given by the following equations [11], where K_{II} is the Mode-II stress intensity factor, μ is the shear modulus, κ is a constant which is $3 - 4\nu$ for PE and $(3 - \nu)/(1 + \nu)$ for PS, and r and θ are polar coordinates measured with respect to the crack tip.

$$u_x = \frac{K_{\text{II}}}{2\mu}\sqrt{\frac{r}{2\pi}}\sin\frac{\theta}{2}\left[\kappa + 1 + 2\cos^2\frac{\theta}{2}\right] \tag{10.1a}$$

$$u_y = \frac{K_{\text{II}}}{2\mu}\sqrt{\frac{r}{2\pi}}\cos\frac{\theta}{2}\left[\kappa - 1 - 2\sin^2\frac{\theta}{2}\right] \tag{10.1b}$$

The above equations can also be arrived at based on the bi-material displacement field equations given in Chapter 5 by setting the "Material 1" and "Material 2" properties to be equal. The above boundary conditions are applied to 26 atomic groups on the boundary of the MD domain (using a boundary domain width of 2 nm), which are shown as a schematic in Figure 10.1.

Loading is applied in increments of $\Delta K_{\text{II}} = 0.1$ MPa$\sqrt{\text{m}}$ over time steps of 2 ps. The constants μ and κ required in Equations (10.1a) and (10.1b) are based on the results from uniaxial tension tests on amorphous carbon shown in Chapter 8 (Table 8.1).

Mode-II MD simulation studies are performed on specimens of varying sizes: (a) 22 × 14 nm with a 12 nm crack, (b) 40 × 24 nm with a 23 nm crack, (c) 76 × 44 nm with a 43 nm crack. The MD domain geometries are selected based on the LEFM fracture process zone sizes and shapes as discussed in Chapter 5.

Fatigue studies on amorphous carbon (a-C) are performed based on the Mode-I specimens described in Chapter 8. Due to the high computational cost of performing repeated cycles, the smallest specimen size, i.e., 14 × 22 nm with a 5 nm crack, is used. For the baseline a-C specimen, crack initiation was found to occur at $K_{\text{I}} = 6$ MPa$\sqrt{\text{m}}$ in Chapter 8. For the fatigue tests, the applied fatigue load was cycled between $K_{\text{I}} = 0$ MPa$\sqrt{\text{m}}$ and $K_{\text{I}} = 5.5$ MPa$\sqrt{\text{m}}$ (immediately prior to crack initiation), giving $\Delta K_{\text{I}} = 5.5$ MPa$\sqrt{\text{m}}$, and the stress ratio $R = 0$. While it is acknowledged that the LEFM fracture process zone (FPZ) size at $K_{\text{I}} = 5.5$ MPa$\sqrt{\text{m}}$ slightly exceeds the MD domain boundaries for the selected small specimen, inaccuracies resulting from this violation of the small-scale yielding (SSY) criterion are not expected to invalidate the qualitative aspects of the fatigue testing results. In the atomistic fatigue studies presented here, it is interesting to track the evolution of damage at the crack tip. Material stiffness degradation is one of the metrics that will be presented in Section 10.3.2.

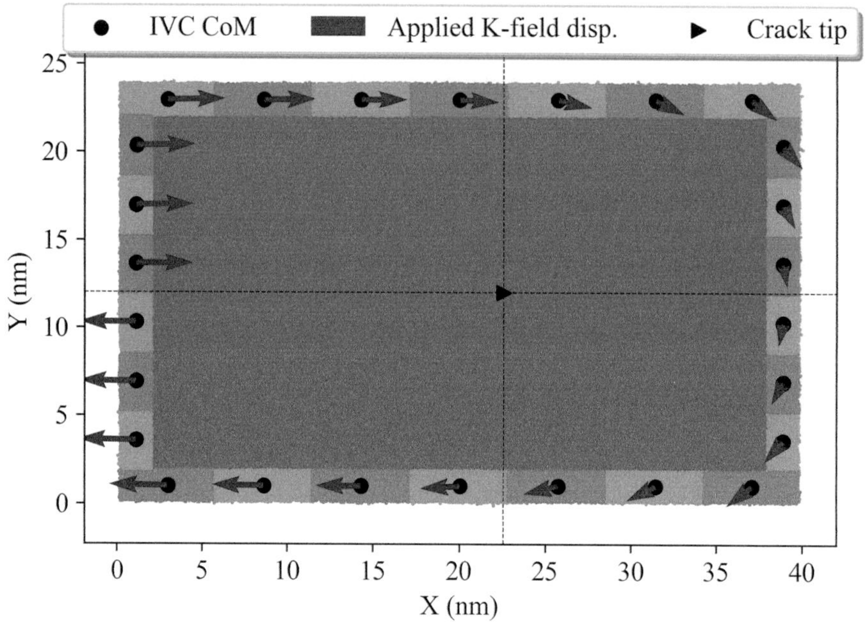

FIGURE 10.1 Mode-II K-test boundary conditions applied to atomic groups on boundary of a cracked 40 × 23 nm MD specimen, with a mid-plane crack of length 23 nm indicated by the arrowhead

To estimate the material stiffness degradation, i.e., degradation in local Young's modulus, stress is measured at the crack tip using a computation volume of radius 1 nm by volume averaging per-atom stress values at the peak of every loading cycle. To estimate strain normal to the crack ε_{yy}, atomistic displacements are tracked 2 nm above and below the crack tip, providing us with l_y and Δl_y so that strain may be calculated at each cycle $\varepsilon_{yy} = \Delta l_y / l_y$. Therefore, using these estimates of crack tip stress and strain, the evolution of material stiffness (i.e., local Young's modulus) with fatigue loading cycles can be plotted.

10.3 RESULTS AND DISCUSSION

10.3.1 Simulation of Mode-II Fracture Loading Using *K*-Test Methodology

Fracture tests are performed with Mode-II loading for the three sizes of MD K-test specimens. Similar to Mode-I fracture, crack initiation in Mode-II is determined based on the peak stress at the crack tip. When the von-Mises (VM) stress at the crack tip reaches a peak value and then begins to fall, it signifies the material has started to fail at the crack tip, and hence, the crack has initiated. A plot of the von-Mises stress at the crack tip varying with applied loading K_{II} for the three MD specimen sizes is shown in Figure 10.2.

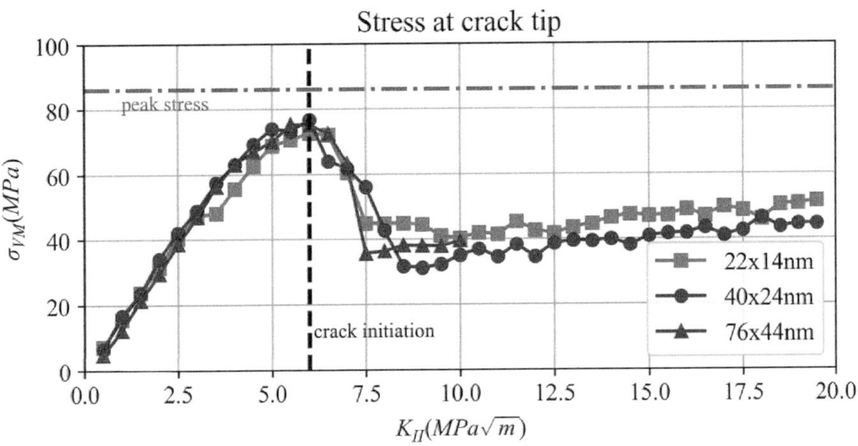

FIGURE 10.2 Computed Von-Mises stress at the crack tip with increasing applied Mode-II loading for various MD specimen sizes.

Based on Figure 10.2, all three MD K-test specimen sizes exhibit similar trends in the peak stress at the crack tip. In all cases, the peak von-Mises stress occurs at $K_{\mathrm{II}} = 6\ \mathrm{MPa}\sqrt{\mathrm{m}}$. Therefore, the initiation fracture toughness in Mode-II based on these results is $K_{\mathrm{II}c} = 6\ \mathrm{MPa}\sqrt{\mathrm{m}}$. To confirm that the K-field loading and von-Mises (VM) stress calculations are performed correctly, we plot the VM stress as a function of distance from the crack tip along the line ahead of the crack tip and compare the MD-computed distribution with the linear elastic fracture mechanics (LEFM) curve. Figure 10.3 shows the variation of VM stress for the three MD specimen sizes along the line immediately prior to crack initiation at $K_{\mathrm{II}} = 5.5\ \mathrm{MPa}\sqrt{\mathrm{m}}$.

As observed in Figure 10.3, the MD-computed von-Mises (VM) stress ahead of the crack tip matches very well with linear elastic fracture mechanics (LEFM) results for all three specimen sizes at crack initiation. Toward the right end of each curve, the values deviate from LEFM, as expected due to domain boundary effects. The LEFM stress fields under Mode-II loading are as follows [11].

$$\sigma_{xx} = -\frac{K_{\mathrm{II}}}{\sqrt{2\pi r}} \sin\frac{\theta}{2}\left[2 + \cos\frac{\theta}{2}\cos\frac{3\theta}{2}\right] \tag{10.2a}$$

$$\sigma_{yy} = \frac{K_{\mathrm{II}}}{\sqrt{2\pi r}} \sin\frac{\theta}{2}\cos\frac{\theta}{2}\cos\frac{3\theta}{2} \tag{10.2b}$$

$$\tau_{xy} = \frac{K_{\mathrm{II}}}{\sqrt{2\pi r}} \cos\frac{\theta}{2}\left[1 - \sin\frac{\theta}{2}\sin\frac{3\theta}{2}\right] \tag{10.2c}$$

Along the line ahead of the crack tip, $\theta = 0$, hence, σ_{xx} and σ_{yy} are zero, and $\tau_{xy} = \frac{K_{\mathrm{II}}}{\sqrt{2\pi r}}$. The out-of-plane component of stress is also zero, i.e. $\sigma_{zz} = 0$. To validate this, the MD computed volumetric stress $\sigma_{\mathrm{vol}} = \frac{1}{3}\left(\sigma_{xx} + \sigma_{yy} + \sigma_{zz}\right)$ is

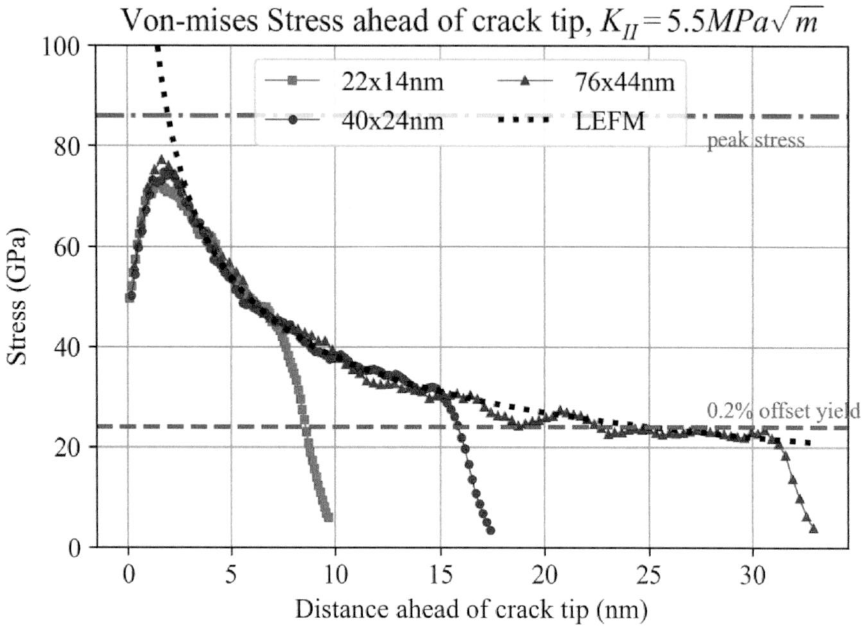

FIGURE 10.3 Von-Mises stress as a function of distance from crack tip along the line ahead of the crack tip for three different MD domain sizes at $K_{\mathrm{II}} = 5.5\ \mathrm{MPa}\sqrt{\mathrm{m}}$.

plotted in Figure 10.4 and, as expected, is approximately zero along the line ahead of the crack tip, thereby further validating the approach.

Finally, a contour plot of the von-Mises stress across the entire MD domain for the largest domain (76 × 44 nm) is shown in Figure 10.5.

The contour line at the yield stress, $\sigma_{\mathrm{yld}} = 22$ GPa, is marked as a solid red line and signifies the spatial extent of the FPZ at $K_{\mathrm{II}} = 5.5\ \mathrm{MPa}\sqrt{\mathrm{m}}$. Also shown is the LEFM-predicted FPZ, marked as a dotted black line. There is excellent agreement between the MD and LEFM-predicted FPZ shapes and sizes. Moreover, at this loading, which is immediately prior to crack initiation, the fully developed FPZ is reasonably well-contained within the MD domain boundaries. Therefore, the application of K-field boundary conditions is validated for Mode-II loading.

Next, the atomistic J-integral is computed for each of the three MD specimen sizes used in this study and plotted against increasing far-field Mode-II load K_{II} in Figure 10.6.

Plotted alongside is the LEFM curve, given by $G = \frac{K_{\mathrm{II}}^2}{E}$, where E is the Young's modulus. It should be noted that Figure 10.6 shows the variation of J_1, which is the total strain energy release rate, i.e., $J_1 = J^{\mathrm{I}} + J^{\mathrm{II}} + J^{\mathrm{III}} = G$, where J^{I}, J^{II}, and J^{III} are the Mode-I, Mode-II, and Mode-III strain energy release rates, respectively. For the pure Mode-II case, $J_1 = J^{\mathrm{II}} = G_{\mathrm{II}}$.

Based on Figure 10.6, the medium (40 × 24 nm) and large (76 × 44 nm) MD domains exhibit J_1 vs. K_{II} curves that match very well with LEFM up to approximately $K_{\mathrm{II}} = 5.5\ \mathrm{MPa}\sqrt{\mathrm{m}}$. These results are consistent with the finding that crack

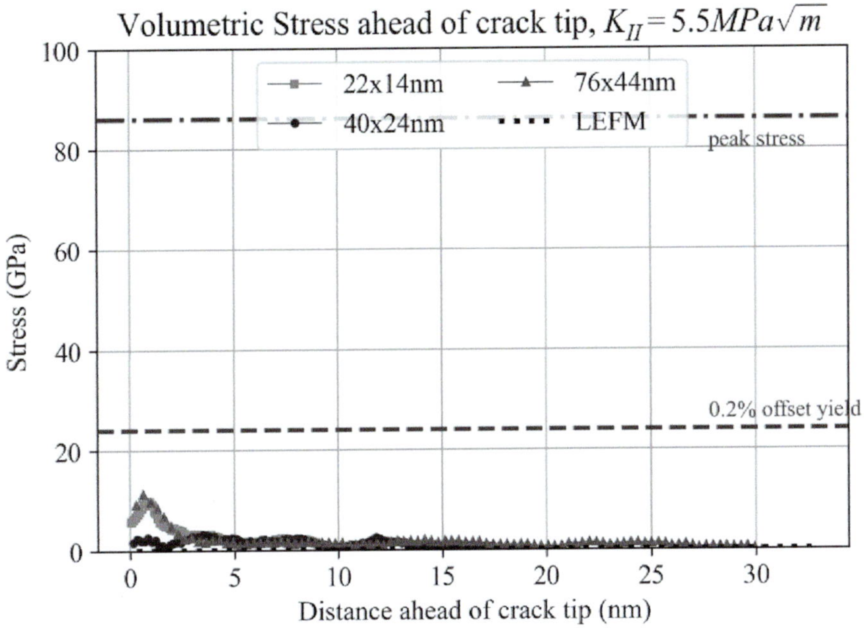

FIGURE 10.4 Volumetric stress as a function of distance from crack tip along the line ahead of the crack tip for various MD specimen sizes at $K_{\text{II}} = 5.5\ \text{MPa}\sqrt{\text{m}}$

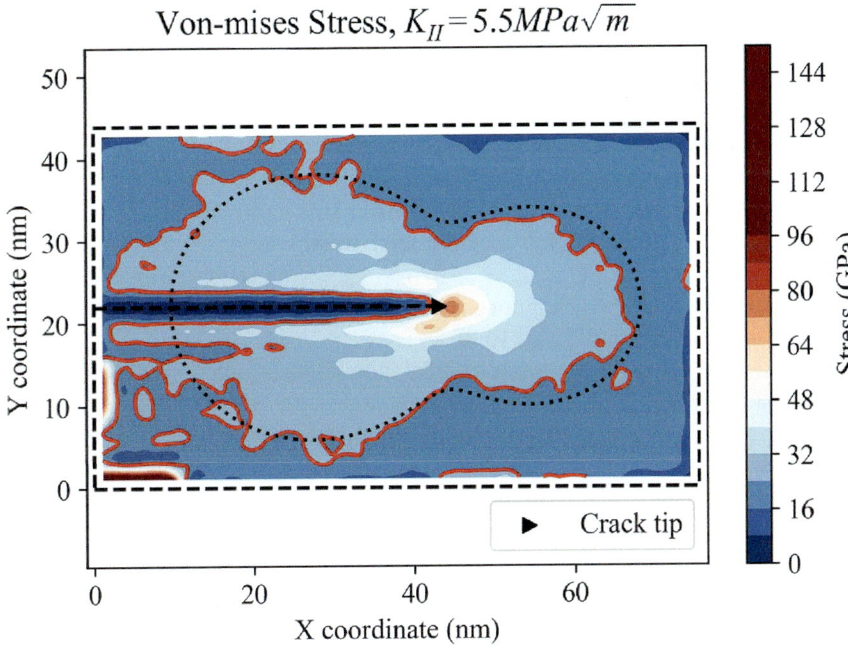

FIGURE 10.5 Von-Mises stress contour plot in 76 × 44 nm specimen with contour lines marked at yield stress to visualize extent of the fully developed FPZ at $T = 300$ K. The solid red line is based on MD stresses, and the black dotted line is based on predictions from LEFM.

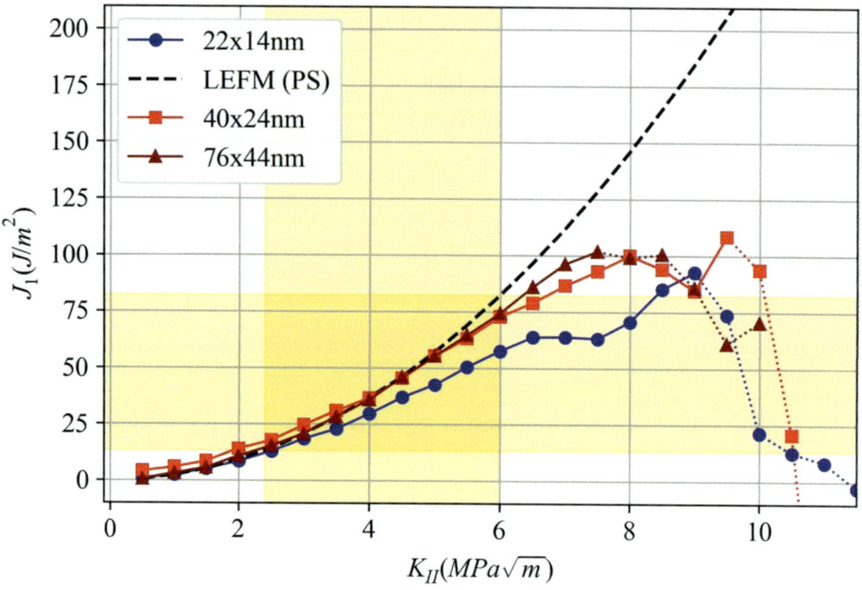

FIGURE 10.6 J_{I} vs. K_{II} for Mode-II loading cases with varying MD domain size. The linear elastic fracture mechanics (LEFM) curve is marked as a dashed black line.

initiation occurred at $K_{\mathrm{II}} = 6\ \mathrm{MPa}\sqrt{\mathrm{m}}$, based on the fact that the peak von-Mises stress occurs at $K_{\mathrm{II}} = 6\ \mathrm{MPa}\sqrt{\mathrm{m}}$ as shown in Figure 10.2. Notably, the J_{I} value for the small MD domain (22 × 14 nm, blue curve in Figure 10.6) begins to deviate from LEFM much earlier and has lower values compared with the other MD specimens. This is expected, since the FPZ is not fully contained by the small domain size at higher loads (greater than $K_{\mathrm{II}} = 3\ \mathrm{MPa}\sqrt{\mathrm{m}}$), and therefore, it violates the K-test rule that the FPZ must be less than the K-test domain size. From Figure 10.6, it can be estimated that the predicted value of the total strain energy release rate at crack initiation in Mode-II for a-C is $G_c = 75\ \mathrm{J/m^2}$.

A visualization of the crack propagation in Mode-II loading of the largest domain (76 × 44 nm) is shown in Figure 10.7.

Color coding is based on atomic coordination number with lower coordination, which indicates lower connections to adjacent atoms, being colored darker and thus indicating development of damage or voids. Figure 10.7 shows the specimen at three loading levels, from crack initiation at $K_{\mathrm{II}} = 6\ \mathrm{MPa}\sqrt{\mathrm{m}}$ to significant crack propagation at $K_{\mathrm{II}} = 10\ \mathrm{MPa}\sqrt{\mathrm{m}}$ when the crack has almost fully propagated through the MD specimen.

The angle at which a crack propagates in Mode-II can be analytically calculated as given in Sih [12], which for a central crack under in-plane shear, is given by the following equation,

$$\theta_0 = -\cos^{-1}\left(\frac{\kappa - 1}{6}\right) \tag{10.3}$$

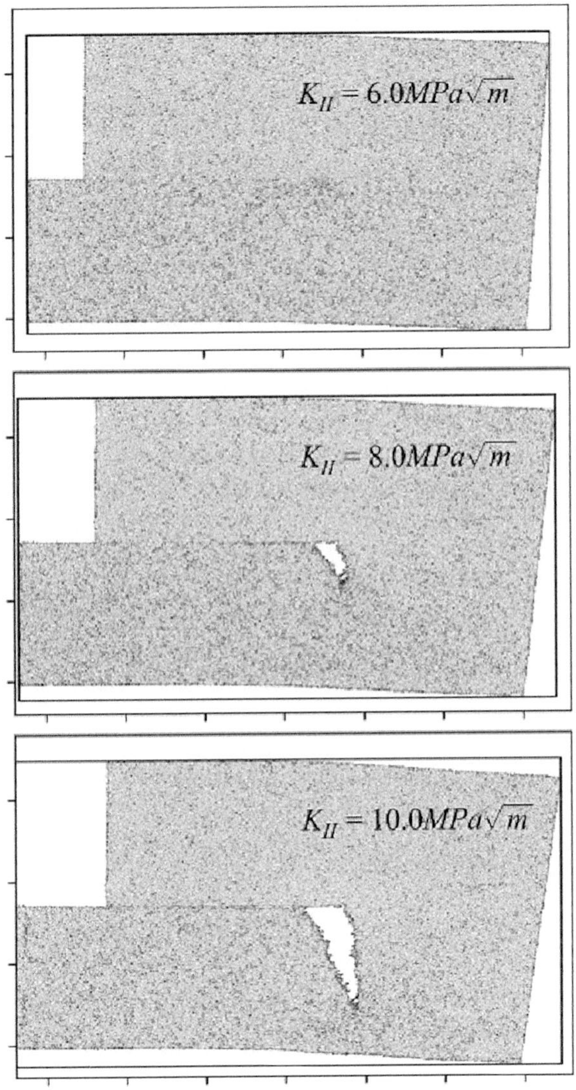

FIGURE 10.7 Visualization of fractured Mode-II specimen of size 76 × 44 nm at $K_{\mathrm{II}} = 6, 8,$ and 10 MPa$\sqrt{\mathrm{m}}$, color coded based on coordination number.

where $\kappa = (3 - \nu)/(1 + \nu)$ for plane stress conditions, with ν being the Poisson's ratio.

For the amorphous carbon material being studied herein, the Poisson's ratio is 0.12, measured based on uniaxial tension tests and presented in Chapter 8 (Table 8.1). Therefore, the analytically computed angle of crack propagation based on Equation 10.3 is $\theta_0 = -75^{\circ}$. From the MD simulation results for the large specimen (76 × 44 nm) as depicted in Figure 6.7, the angle of crack propagation is approximately $\theta_0 = -76^{\circ}$, which is in close agreement with the analytical value.

10.3.2 Simulation of Mode-I Fatigue Loading Using *K*-Test Boundary Conditions

Based on Mode-I loading results for amorphous carbon in Chapter 8, crack initiation was observed to occur at $K_1 = 6\ \text{MPa}\sqrt{\text{m}}$. Therefore, for the MD-based fatigue test simulations in this section, Mode-I loading is applied with $\Delta K_1 = 5.5\ \text{MPa}\sqrt{\text{m}}$ to reach fatigue loading level approximately 9% below the static crack initiation level. For the fatigue test simulations, the applied fatigue load was cycled between $K_1 = 0\ \text{MPa}\sqrt{\text{m}}$ and $K_1 = 5.5\ \text{MPa}\sqrt{\text{m}}$ (immediately prior to crack initiation), giving $\Delta K_1 = 5.5\ \text{MPa}\sqrt{\text{m}}$, with a stress ratio of $R = 0$. The smallest MD specimen size studied in Chapter 8, i.e., 14 × 22 nm with a 5 nm crack, is used in this fatigue study, primarily to allow for higher computational efficiency while performing numerous cycles of loading. The MD specimen for fatigue test simulation is identical to that for the Mode-I *K*-test simulation for a-C presented in Chapter 8 and is therefore not separately depicted here. Due to the extremely high computational cost of running *K*-test-based fatigue simulations, results for a single MD realization of the fatigue specimen are presented here, with the understanding that for statistically meaningful results, multiple MD realizations need to be included. The cyclic loading rate used in the Mode-I fatigue simulation is approximately 4.5/ns, which is very high compared with real-life fatigue loading rates. The applied loading is incremented (or decremented) in steps of $0.1\ \text{MPa}\sqrt{\text{m}}$. At each loading level, the MD system is equilibrated over 2 ps of simulation time, which was found to be long enough to allow the system potential energy to converge. Therefore, this method results in quasi-static Mode-I fatigue loading.

The computed stresses at the crack tip are found to decrease with the number of cycles, indicating material degradation in the crack tip region. In Figure 10.8, the approximate material stiffness at the crack tip is plotted as a function of fatigue loading cycles up to ~600 cycles.

The stiffness (or modulus) is measured using the ratio of the stress to the strain at peak applied load ($K_1 = 5.5\ \text{MPa}\sqrt{\text{m}}$). The stiffness vs. number of cycles plot in Figure 10.8 indicates the occurrence of multiple nanoscale failure events resulting in around 20% net drop in stiffness, from 300 to 240 GPa, over 600 cycles. It should be noted that the stiffness measure presented here approximately quantifies the material stiffness of the region within ~2 nm radius of the crack tip. The gradual degradation of stiffness at the crack tip, as shown in Figure 10.8, is indicative of an evolving inelastic process zone at the crack tip. This inelasticity can be quantified by observing the residual strain at the crack tip at the end of every loading cycle, as plotted in Figure 10.9.

The residual strain plot starts from ~0.5% strain, which is the value after a single loading cycle, and rises up to ~1.2% after 600 cycles. For both the stiffness and the residual strain, the rate of change starts to diminish with increasing loading, as is expected for displacement-controlled loading. Based on Figures 10.8 and 10.9, there are significant nanoscale failure events that occur right at the start of fatigue loading (between 1 and 4 cycles) and then again after around 100 cycles. To confirm that visible damage has evolved for the MD model under fatigue loading, a visualization of the near crack tip region in the MD domain is shown in Figure 10.10 for up to 600 cycles. The vertical black line indicates the position of the original crack tip and is

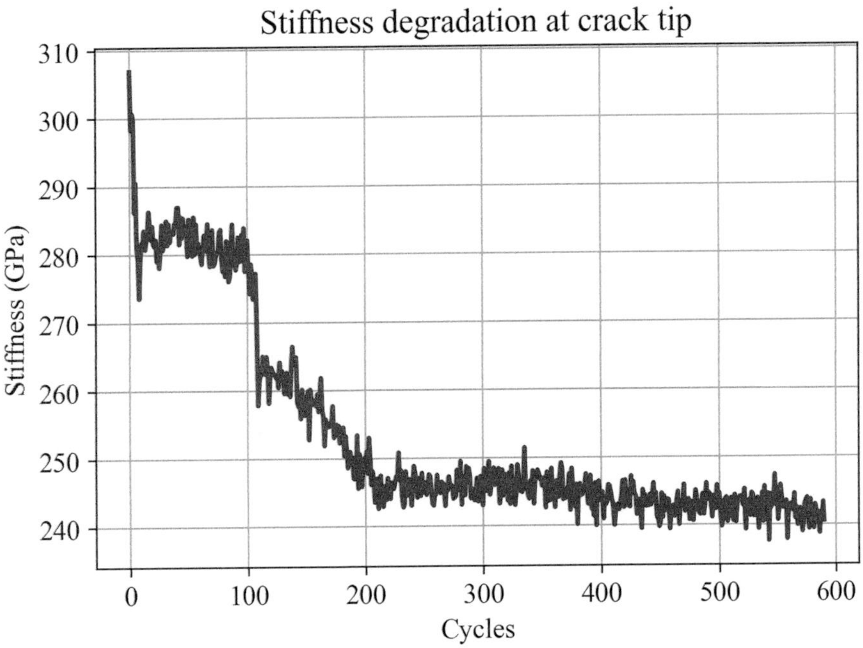

FIGURE 10.8 Variation of stiffness (modulus) near the crack tip, measured using stress and strain at peak applied load ($K_I = 5.5$ MPa$\sqrt{\text{m}}$), plotted against fatigue cycles at $T = 300$ K.

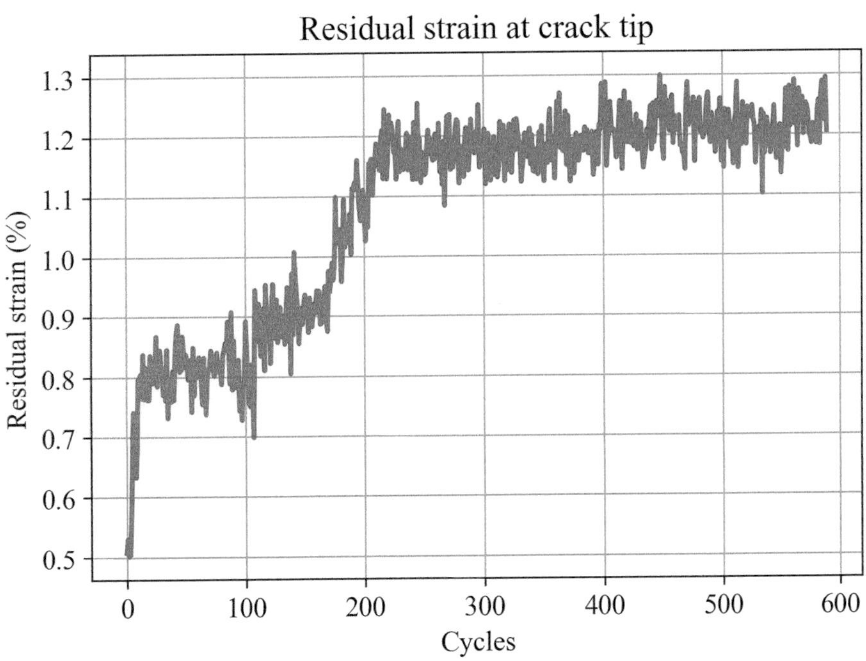

FIGURE 10.9 Variation of residual strain near the crack tip at peak applied load ($K_I = 5.5$ MPa$\sqrt{\text{m}}$) with fatigue cycles at $T = 300$ K.

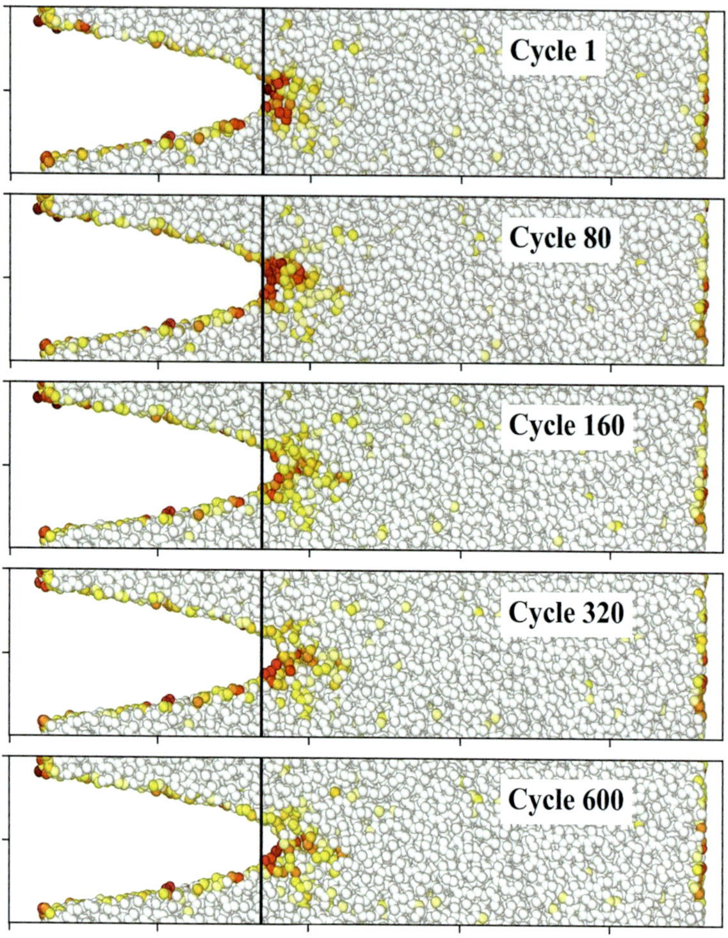

FIGURE 10.10 Visualization of near crack tip region in MD domain at cycle number 1, 80, 160, 320, and 600 at $T = 300$ K. Color coding is based on coordination number with darker red indicating lower atomic coordination number. The vertical black line is used as reference to indicate the original position of the crack tip.

used as reference for crack growth. It is likely that the sudden drop in the stiffness observed in Figure 10.8 is due to fatigue crack initiation after 100 cycles of loading. Nanoscale crack growth can be clearly seen in Figure 10.10, thus confirming that localized failure events in the FPZ occur during the fatigue cycles.

The fatigue crack growth is accompanied by a reduction in atomic coordination number. Lower coordination number is shown as darker red in Figure 10.10, indicating fewer covalent bonds between atoms near the crack tip due to fatigue damage. As can be seen from Figures 10.8 and 10.10, fatigue crack growth rate and damage show significant slowdown beyond 200 cycles.

10.3.3 Extraction of Paris Law Parameters from MD Data

As a case study for the Mode-I MD K-test fatigue analysis, we benchmark how well the MD-based crack extension results compare with established macro-scale fatigue models, such as the Paris law, and the feasibility of extracting Paris law parameters from MD data. As depicted in Figure 10.10, there is a measurable, albeit small, extension of the crack at the nanoscale. To accurately pinpoint the location of the crack tip, the density of the material along the crack line is used. Moving from left to right, the density jumps upon crossing the crack tip, thereby giving us the exact location of the crack tip and the crack extension due to fatigue cycling. Figure 10.11 shows a plot of the crack extension vs. number of fatigue cycles, with the MD data points shown as solid gray circles.

To help understand the relationship between the crack extension and the number of cycles in the present test, a power law fit to the MD data is shown as a solid blue line in Figure 10.11.

As expected, the crack rapidly extends in the initial cycles and then the rate of crack extension reduces. This behavior is consistent with the style of loading that is employed in this case, wherein displacement fields are prescribed at the MD domain (K-test) boundaries, as opposed to far-field load control. Typically, in displacement-controlled fatigue tests, the material compliance increases due to local damage formation at the crack tip, and therefore, the stress intensity decreases with the number of cycles. The fact that the material degrades and becomes more compliant

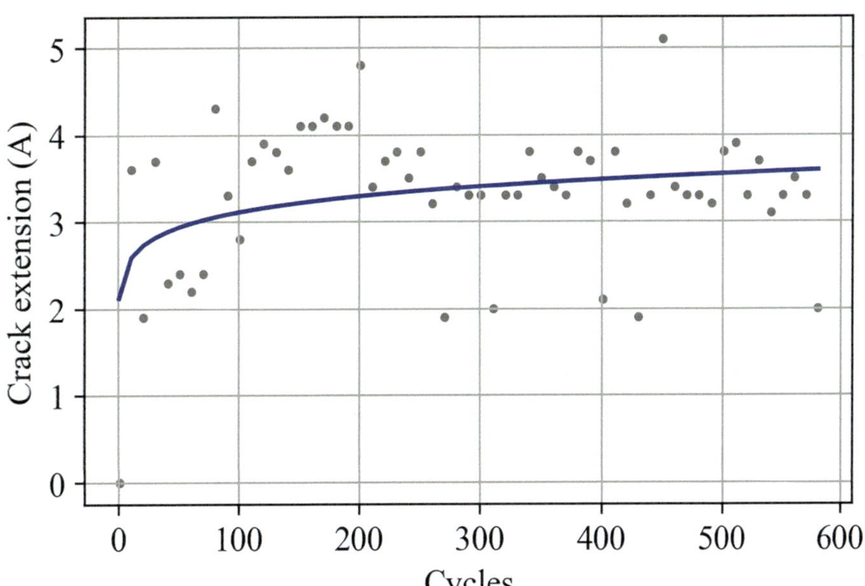

FIGURE 10.11 Crack extension vs. cycles for fatigue loading on amorphous carbon using K-test MD simulations at $T = 300$ K. Gray circles are MD data points. The blue line is a power law fit.

is evident from the modulus degradation depicted in Figure 10.8. In attempting to fit a fatigue crack extension model to the MD data, it is essential, therefore, to account for this tensile modulus degradation, and hence, relaxation in stress intensity near the vicinity of the crack. A simple way to estimate the relaxation of the cyclic stress intensity factor range (ΔK) is by using the K-dominant displacement field equations given in Chapter 5. From Equations (5.1a) and (5.1b), it can be deduced that for the applied K-test boundary displacement fields u_1 and u_2 to remain constant with increasing number of cycles, the ratio of stress intensity factor to the shear modulus $\left(\frac{K_1}{\mu}\right)$ must remain essentially constant throughout the fatigue loading. Assuming that degradation in the Poisson's ratio is negligible, this implies that the reduction in K_I is directly proportional to the reduction in elastic modulus near the crack tip (i.e., $K_1 \propto E$), as depicted in Figure 10.8. Therefore, ΔK, which has an initial value of $\Delta K = 5.5 \text{ MPa}\sqrt{\text{m}}$, is allowed to decrease proportionately with the tensile modulus to a value of $\Delta K = 4.4 \text{ MPa}\sqrt{\text{m}}$ at the end of 600 cycles. While more accurate methods exist to estimate the stress intensity factor range, the present method provides a reasonably good estimate.

As mentioned in Chapter 1, in the stable crack growth regime, Paris law takes the following form

$$\frac{da}{dN} = A(\Delta K)^m \tag{10.4a}$$

where a is the crack length, N is the number of cycles, ΔK is the cyclic stress intensity factor range, and A and m are Paris law parameters which are material constants.

By taking logarithm on both sides of Equation (10.4a), it takes the linearized form:

$$\log \frac{da}{dN} = \log C + m \log (\Delta K) \tag{10.4b}$$

The above equation can now be fitted to the MD data points via linear regression, and the results are shown in Figure 10.12.

While there is some scatter in the MD data points, overall, a clear linear trend is observed. Linear regression fit to the data in Figure 10.12 yields $m = 14.5$ and $\log A = -29$. The linear fit has a correlation coefficient $R^2 = 0.93$, which indicates that it is a reasonably good fit. Having obtained values for m and A, it is now feasible to predict the number of cycles to complete ligament failure (N_f) for the MD specimen using Equation (1.16), albeit the high value of m indicates that the obtained fatigue life will be relatively short for this demonstration case, due to length and time scale effects in our MD simulation. This underscores the need for developing proper scaling laws to bridge length and time scales, similar to the one presented in Al Amin et al. [13]. Nevertheless, the fatigue simulation framework, presented herein, presents pathways for the prediction of Paris law parameters that can potentially be employed for fatigue life estimation at the macro-scale, especially if the material response is independent of the cyclic loading rate.

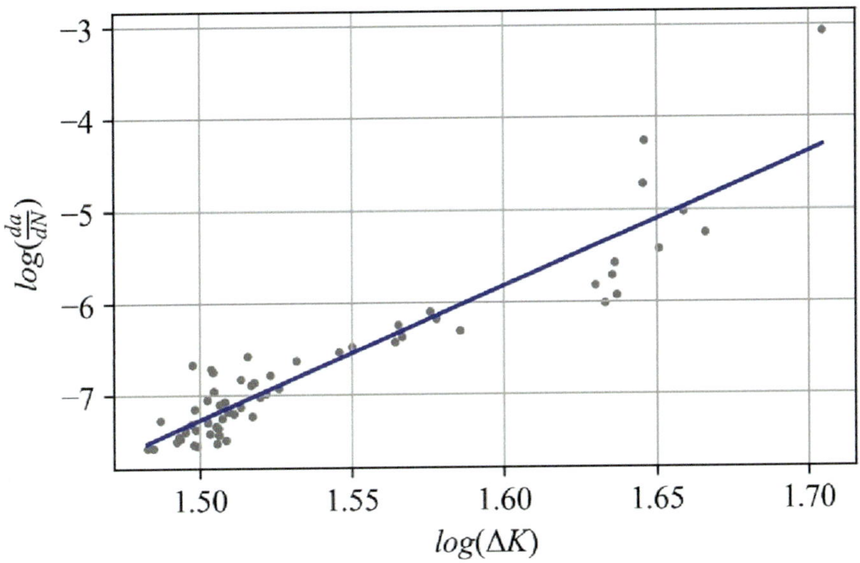

FIGURE 10.12 Log-log plot of the rate of crack extension vs. cyclic stress intensity factor range for fatigue loading on amorphous carbon using Mode-I K-test MD simulations at $T = 300$ K. Gray circles are MD data points. The blue line is a linear fit to Paris law.

10.4 SUMMARY

This chapter consists of two distinct parts, both related to amorphous carbon (a-C). The first part described Mode-II fracture loading in a-C using three different specimen sizes, based on the expected shapes of the fracture process zone (FPZ) from linear elastic fracture mechanics (LEFM). Equations and K-test methodology for Mode-II were laid out, with many aspects being similar to Mode-I loading discussed in Chapter 8. The crack initiation criterion for the Mode-II case was based on the peak von-Mises stress at the crack tip. Various plots of stress, as computed from the MD simulations, were presented and compared with LEFM predictions, showing good agreement and therefore validating the approach for Mode-II. The atomistic J-integral vs. applied K_{II} also showed good agreement with LEFM up to the point of crack initiation, as is expected. At crack initiation, the MD-predicted values for Mode-II toughness for a-C are $G_c = 75\ \text{J/m}^2$ and $K_{\text{II}c} = 6\ \text{MPa}\sqrt{\text{m}}$. While experimental data on Mode-II fracture toughness of amorphous carbon could not be found in the literature, the $G_c = 75\ \text{J/m}^2$ value measured using MD Mode-II tests matched well with MD Mode-I tests in Chapter 8, as well as with experimentally measured Mode-I ranges.

The second part of the chapter deals with Mode-I fatigue simulations for amorphous carbon based on the K-test methodology used for fracture testing, applied in a cyclic manner. More than 600 cycles were performed on the a-C specimen under Mode-I loading, up to a point just prior to crack initiation in standard Mode-I fracture tests, showing a steady decline in modulus indicative of gradual material degradation. Local damage formation and fatigue crack growth were observed both visually, as well as based on stiffness degradation and development of local residual strain. Paris law parameters were extracted from the MD data and established the feasibility for potentially predicting macro-scale fatigue life of a material using crack growth data from MD simulations. Therefore, the applicability of the K-test methodology for simulating fatigue crack growth in amorphous materials was established in this chapter.

Overall, this chapter presented novel methodologies for Mode-II fracture and Mode-I fatigue loading using the steered MD-based K-test framework at room temperature $(T = 300\text{ K})$. To our knowledge, a mechanistically sound analysis of these types of loading has not been performed in the literature. While this was a preliminary study of Mode-II fracture and Mode-I fatigue loading in amorphous carbon, future studies can be designed to examine the effect of nanoparticles on crack tip shielding effects for these loading cases, analogous to the Mode-I studies presented in Chapter 9.

REFERENCES

1. Wei, Y., et al., Fracture properties of nanoscale single-crystal silicon plates: Molecular dynamics simulations and finite element method. *Engineering Fracture Mechanics*, 2018. **202**: p. 1–19.
2. Dontsova, E. and R. Ballarini, Atomistic modeling of the fracture toughness of silicon and silicon-silicon interfaces. *International Journal of Fracture*, 2017. **207**: p. 99–122.
3. Tabarraei, A. and X. Wang, A molecular dynamics study of nanofracture in monolayer boron nitride. *Materials Science and Engineering: A*, 2015. **641**: p. 225–230.
4. Zhou, X., et al., Molecular dynamics simulation based cohesive surface representation of mixed mode fracture. *Mechanics of Materials*, 2008. **40**(10): p. 832–845.
5. Tang, T., S. Kim, and M. Horstemeyer, Fatigue crack growth in magnesium single crystals under cyclic loading: Molecular dynamics simulation. *Computational Materials Science*, 2010. **48**(2): p. 426–439.
6. Ma, L., et al., Molecular dynamics simulation of fatigue crack propagation in bcc iron under cyclic loading. *International Journal of Fatigue*, 2014. **68**: p. 253–259.
7. Wu, W.-P., Y.-L. Li, and X.-Y. Sun, Molecular dynamics simulation-based cohesive zone representation of fatigue crack growth in a single crystal nickel. *Computational Materials Science*, 2015. **109**: p. 66–75.
8. Liu, X., et al., Molecular dynamics simulation of effects of loading parameters on fatigue crack growth behavior in titanium single crystal. *Fatigue & Fracture of Engineering Materials & Structures*, 2024. **47**(5): p. 1715–1730.
9. Holopainen, S. and T. Barriere, Modeling of mechanical behavior of amorphous solids undergoing fatigue loadings, with application to polymers. *Computers & Structures*, 2018. **199**: p. 57–73.
10. Yue, T., et al., Fatigue Behavior of Polymer Nanocomposites under Low-Strain Cyclic Loading: Insights from Molecular Dynamics Simulation. *Langmuir*, 2024. **40**(45): p. 23816–23824.
11. Wang, C.H., *Introduction to fracture mechanics*. 1996: DSTO Aeronautical and Maritime Research Laboratory Melbourne, Australia.
12. Sih, G.C., Strain-energy-density factor applied to mixed mode crack problems. *International Journal of Fracture*, 1974. **10**: p. 305–321.
13. Al Amin, M.D. and S. Roy, Atomistic simulation of damage in carbon-carbon composite from impact with water droplets at hypersonic velocity. *Carbon*, 2025. **245**: p. 120729.

Index

Pages in *italics* refer to figures and pages in **bold** refer to tables.

L

M

N

T

U

V

Y

For Product Safety Concerns and Information please contact our
EU representative GPSR@taylorandfrancis.com Taylor & Francis
Verlag GmbH, Kaufingerstraße 24, 80331 München, Germany